DEVELOPING AREAS RESEARCH GROUP
THE ROYAL GEOGRAPHICAL SOCIETY
(WITH THE INSTITUTE OF BRITISH GEOGRAPHERS)

DARG Regional Development Series No. 2
Series Editor: David Simon

Challenges and Change in Middle America

Perspectives on development in Mexico, Central America and the Caribbean

edited by Cathy McIlwaine and Katie Willis

Longman

An imprint of **Pearson Education**

Harlow, England · London · New York · Reading, Massachusetts · San Francisco
Toronto · Don Mills, Ontario · Sydney · Tokyo · Singapore · Hong Kong · Seoul
Taipei · Cape Town · Madrid · Mexico City · Amsterdam · Munich · Paris · Milan

Pearson Education Limited
Edinburgh Gate
Harlow
Essex CM20 2JE
and Associated Companies throughout the world.

Visit us on the World Wide Web at:
www.pearsoneduc.com

ISBN 0582-40485-1

British Library Cataloguing-in-Publication Data
A catalogue record for this book is available from the British Library

Library of Congress Cataloging-in-Publication Data
Challenges and change in Middle America : perspectives on development in Mexico,
Central America and the Caribbean / edited by Cathy McIlwaine and Katie Willis.
    p. cm. — (DARG regional development series ; no. 2)
    Includes bibliographical references and index.
    ISBN 0-582-40485-1
    1. Central America—Economic conditions—1979– 2. Caribbean Area—Economic
conditions—1945– 3. Mexico—Economic conditions—1994– I. McIlwaine, Cathy, 1965–
II. Willis, Katie, 1968– III. Series.

HC141 .C42 2001
338.972—dc21   2001036586

10 9 8 7 6 5 4 3 2 1
05 04 03 02

Typeset in $^{10}/_{11}$ Palatino by 35
Printed and bound by Malaysia , LSP

In memory of Derek McIlwaine

Contents

vii

Contents

Plates

Figures

Tables

Tables

Boxes

Boxes

Contributors

Jonathan Barton is a Lecturer in Environmental Policy at the School of Development Studies, University of East Anglia. His interests include the effectiveness of regulatory systems and environment agencies, business responses to environmental policy, and broader political ecology themes relating to economic development in Latin America. Recent research has included work on aquaculture and environment issues in Chile, and the impacts of environmental regulations on the iron and steel sectors in the European Union (EU), Republic of Korea, Czech Republic, Poland and Brazil. Publications from this research include: 'Environment, Sustainability and Regulation in Commercial Aquaculture: the case of Chilean salmonid production' (*Geoforum* 1998), 'The North–South Dimension of the Environmental and Cleaner Technology Industries' (*CEPAL Review* 1998) and 'The Environmental Agenda: accountability for sustainability' in N. Phillips and J. Buxton (eds) *Developments in Latin American Political Economy* (Manchester University Press 1999). He is also author of *A Political Geography of Latin America* (Routledge 1997).

Elizabeth Bennett is a natural resources economist in the Centre for the Economics and Management of Aquatic Resources, University of Portsmouth. She graduated with a first degree in Latin American Studies and has since gone on to specialise in natural resource management in developing countries. She lectures on Latin American economics, globalisation issues and fisheries management and development. She is currently registered for a PhD in Economics. Recent publications include 'Saint Mary of the Contra, Our Lady of Oliver North: Catholicism and the Nicaraguan State' (*International Journal of Social Economics* 2000); 'Compliance and fisheries governance: the importance of legitimacy' (*Land Economics* 2000) and 'Globalisation and the sustainability of world fisheries: a view from Latin America' (University of Portsmouth 2000).

Sarah Bradshaw is a Lecturer in Gender and Development at Middlesex University, and is currently on a two-year sabbatical working in Nicaragua. She works with a Nicaraguan feminist non-governmental organisation (NGO) – *Fundación Puntos de Encuentro para Transformar la Vida Cotidiana* – as part of external co-operation through the British NGO, International

Co-operation in Development–Catholic Institute for International Relations (ICD-CIIR).

Sylvia Chant is a Reader in Geography at the London School of Economics and Political Science, with area interests in Latin America (Costa Rica and Mexico), and Southeast Asia (the Philippines). She has published widely in the field of gender and development, female employment, household survival strategies, lone motherhood, and more recently men and masculinities. Her latest books include *Women-headed Households: Diversity and Dynamics in the Developing World* (Macmillan 1997) and *Three Generations, Two Genders, One World: Women and Men in a Changing Century* (with Cathy McIlwaine) (Zed 1998). She is currently working on a book entitled *Gender in Latin America* with Nikki Craske for the Latin America Bureau, and is furthering her research on household and family transformations at the turn of the century.

David Howard is a Lecturer in Human Geography at the University of Edinburgh. His research focuses on the social and urban geographies of the Caribbean, with specific interests in issues of racism, contemporary representations of race and nation and the influence of migration on changing notions of ethnicity. He has most recently completed research projects in Jamaica and the Dominican Republic.

Brian Linneker is a research economist and geographer. Over the last 10 years he has worked as a British Government and EU research consultant as well as being a research officer and lecturer at the London School of Economics and Political Science. He currently lives in Managua and works for the Civil Co-ordinator for Emergency and Reconstruction (CCER) in Nicaragua as part of external co-operation through the British NGO, International Co-operation in Development–Catholic Institute for International Relations (ICD-CIIR). He works with the CCER National Commission for Information and Evaluation of Hurricane Mitch, the Commission for the Social Audit, which monitors and evaluates the emergency and reconstruction efforts, and the CCER Co-ordination Commission.

Sally Lloyd Evans is a Lecturer in Geography at the University of Reading. Her principal research interests are work, employment and the informal sector in both Caribbean and UK labour markets; gender, race and ethnicity; and cities and urban development. Her recent publications include *The City in the Developing World* (with Rob Potter) (Addison Wesley Longman 1998) and *Resource Sustainability and Caribbean Development* (edited with Duncan McGregor and David Barker) (University of West Indies Press 1998). Sally is currently working on a research project on racialised gendering in the UK labour market, funded under the Economic and Social Research Council (ESRC) Cities Programme, but is planning to return to the Caribbean in the near future.

Cathy McIlwaine is a Lecturer in Human Geography at Queen Mary, University of London. Her areas of interest include Costa Rica, El Salvador,

Guatemala and Colombia in Latin America, and the Philippines in Southeast Asia. Her research interests include gender, households, employment, civil society and urban violence. She is currently working on a project on urban violence in Colombia and Guatemala funded by the World Bank and the Swedish International Development Co-operation Agency (SIDA). Her recent publications include *Three Generations, Two Genders, One World: Women and Men in a Changing Century* (with Sylvia Chant) (Zed 1998), *Urban Poor Perceptions of Violence and Exclusion in Colombia* (with Caroline Moser) (World Bank 2000), and *Violence in a Post-Conflict Context: Urban Poor Perceptions from Guatemala* (with Caroline Moser) (World Bank 2000).

Mark Pelling is a Lecturer in Human Geography at the University of Liverpool. He has also worked in the Department of Geography, University of Guyana, and has conducted research into urban sustainability in Guyana, Barbados and the Dominican Republic, as well as working as a consultant in urban poverty for the UK Department for International Development and Government of Barbados. His publications include works in the *Journal of International Development*, *Environment and Urbanisation*, *Geoforum* and *Caribbean Geography*. He is currently writing a book entitled *Cities and Disasters* based on his experience in this region.

Andy Thorpe is a Principal Lecturer in the Department of Economics at the University of Portsmouth. His interest in Central American agrarian issues was aroused during time spent as a Visiting Professor in Agricultural Economics at the Postgrado en Economía y Planificación (POSCAE), National State University of Honduras during the early 1990s. Recent publications include 'The New Economic Model and Fisheries Development in Latin America' (*World Development* 2000), 'It Could be You! Rural Conscription Processes in Honduras' (*European Review of Latin American and Caribbean Studies* 2000) and 'The Power of Policy and the Policy of Power: the Economics of Energy Production and Provision in Honduras' (*Energy and Development* 2000).

Katie Willis is a Lecturer in the Geography Department at the University of Liverpool. Her main research interests are in gender, households and migration. She has published work on gender, class and households in Oaxaca, Mexico, in journals such as the *Bulletin of Latin American Research* and *European Review of Latin American and Caribbean Studies*. Her current research is on skilled migration from the UK and Singapore to China. Recent publications include *Gender and Migration* (edited with Brenda Yeoh) (Edward Elgar 2000) and papers in *Gender, Place and Culture* and *Regional Studies*.

Rebeca E. Zúniga is a sociologist. She worked for two years in Guatemala as a researcher on projects related to human rights violations committed during the civil war, including the Interdiocesan Recovery of Historical Memory (REMHI) project and the Commission for Historical Clarification (CEH). She currently lives in Managua, Nicaragua, and works for the Civil Co-ordinator for Emergency and Reconstruction (CCER) with the CCER National Commission for Information and Evaluation and CCER Co-ordination Commission, Proposal and Strategic Commissions.

Series Preface

In the late 1980s, the Developing Areas Research Group (DARG) of the then Institute of British Geographers produced a series of three edited student texts under the general editorship of Prof. Denis Dwyer (University of Keele). These volumes, focusing on Latin America, Asia and Tropical Africa respectively, were published by Longman and achieved wide circulation, thereby also contributing to the financial security of DARG and enabling it to expand its range of activities. These are now out of print. The Latin American volume was revised and published in a second edition in 1996.

However, there have been dramatic changes in the global political economy, in the nature of development challenges facing individual developing countries and regions, and in debates on development theory over the last decade or so. In a review of the situation in 1997/8, the DARG Committee and Matthew Smith, the Senior Acquisitions Editor at Addison Wesley Longman (now Pearson Education) therefore felt that mere updating of the existing texts would not do these circumstances justice or catch the imagination of a new generation of students. Accordingly, we have launched an entirely new series.

This is both different in conception and larger, enabling us to address smaller, more coherent continental or subcontinental regions in greater depth. The organising principles of the current series are that the volumes should be thematic and issue-based rather than having a traditional sectoral focus, and that each volume should integrate perspectives on development theory and practice. The objective is to ensure topicality and clear coherence of the series, while permitting sufficient flexibility for the editors and contributors to each volume to highlight regional specificities and their own interests. Another important innovation is that the series was launched in January 1999 by a book devoted entirely to provocative contemporary analyses of *Development as Theory and Practice; current perspectives on development and development co-operation*. Edited by David Simon and Anders Närman, this provides a unifying foundation for the regionally focused texts and is designed for use in conjunction with one or more of the regional volumes.

The complete series is expected to include titles on Central America and the Caribbean, Southern and East Africa, West Africa, the Middle East and

North Africa, South Asia, Pacific Asia, the transitional economies of central Asia, and Latin America. While the editors and many contributors are DARG members, other expertise – not least from within the respective regions – is being specifically included to provide more diverse perspectives and representativeness. Once again, DARG is benefiting substantially from the royalties. In addition, a generous number of copies of each volume will be supplied to impoverished higher education institutions in developing countries in exchange for their departmental publications, thereby contributing in a small way to overcoming one pernicious effect of the debt crisis, namely the dearth of new imported literature available to staff and students in those countries.

David Simon
Royal Holloway, University of London

Series Editor
(Chair of DARG 1996–8)

Acknowledgements

Since its inception, this book has been a collaborative effort on the part of a host of different people, not least the contributors themselves. In particular, David Simon, as the editor of the DARG series, has enthusiastically encouraged us through the trials and tribulations of editing the book. Indeed, we would like to thank him warmly, not only for spurring us on, but also for reading the manuscript and providing us with some excellent comments. We would also like to thank Matthew Smith at Pearson Education who commissioned the book and who has also been an enthusiastic supporter throughout the process. Sandra Mather at the Department of Geography, University of Liverpool, drew the maps in chapters 1 and 2 as well as figures in chapters 9 and 10. We are grateful to her for this. We would both like to thank the administrative staff at both Departments of Geography at the University of Liverpool and at Queen Mary, University of London, for helping out at various stages (especially Ed Oliver who helped negotiate the printing of maps and figures).

Personally, this has also been a fruitful and enjoyable experience for us both. Despite this being the first time we have worked together, we have done so effortlessly and we are grateful for the opportunity. Finally, we would like to thank friends and family for giving us additional and much-appreciated support at various stages, especially Lee Drabwell, Janet McIlwaine, Sylvia Chant, Hazel Johnstone and Caroline Sutton, friends in Geography at Queen Mary, all on the fourth floor of the Roxby Building, Kath Owen, Caroline Logan, Bernie O'Neill, Becci Wake and Kuno van der Post.

Publisher's Acknowledgements

We are grateful to the following for permission to reproduce copyright material:

Table 1.1 from *World Bank and UNDP*, Human Development Report 2000, published by Oxford University Press, New York; Table 1.2 from *World Bank and UNDP*, Human Development Report 1999, published by Oxford University Press, New York; Tables 2.1 and 6.1 from *UNDP 1997, World Bank 1995/6*, published by Oxford University Press, New York; Table 2.2 from *UN 1995, UNDP 1995*, published by Oxford University Press, New York; Table 4.1 from *World Bank*, World Development Report 1988, published by Oxford University Press, New York; Table 4.2 from *UNDP*, Human Development Report 1999, published by Oxford University Press, New York; Table 5.3 from *World Bank*, World Development Report 2000, published by Oxford University Press, New York; Table 8.2 from *World Resources Institute 1998*, published by Oxford University Press, New York; Table 1.3 from *EIU 1998a, 1998b, 1999a, 1999b, 2000a, 2000b, 2000c*, Trade Reliance on the United States, published by Economist Intelligence Unit, London; Tables 5.1 and 5.2 from *United National 1995*, The World's Women 1995, published by United Nations, New York; Table 6.2 from *Maquila Portal 2000*, INEGI, Ciemez-Vefa, published by Maquilaportal; Table 7.1 from *Agriculture as a source of regional employment, export earnings and contributor to GDP* (ECLAC, 1999), published by the Food and Agriculture Organisation of the United Nations; Tables 7.2 and 7.4 from *The Political Economy of Central America since 1920* (Bulmer-Thomas, 1994 and FAO, 1999), published by Cambridge University Press, Cambridge; Table 7.3 from *Coffee and Power: Revolution and the Rise of Democracy in Central America* adapted from Paige, J.M., adapted from Harvard University Press, Massachusetts; Table 7.5 from *Impact of agrarian reform* (Baumeister, 1992); Figure 7.1 from *Non-traditional exports from Mexico, Central America and large Caribbean Islands 1980–1998* (FAO data), published by the Food and Agricultural Organisation of the United Nations; Tables 8.1 and 8.5 from *Health in the Americas 1998*, published by the Pan-American Health Organisation. 'To obtain information about Health in the Americas, 1998 Edition, visit their website at http://publications.paho.org or write to the Pan-American Health Organisation, Publications Programme, 525 Twenty-third Street, NW Washington

Publisher's Acknowledgements

DC, 20037, Fax: (202) 338-0869'; Tables 8.3 and 8.4 from *World Resources Institute: World Resources 1992–1993, World Bank 1999*, published by World Resources Institute, Washington; Figures 8.1 and 8.2 from *World Development Indicators 1999*, published by the World Bank, Washington; Figure 9.3 from *Third World Planning Review* (Drakakis-Smith, 1996), published by Routledge, London; Table 10.1 from *Gobierno de Nicaragua 1999, 2000a and CCER*, Authors' estimate from official and unofficial sources, published by the World Bank, Washington.

Plates 9.1 and 9.2 reproduced by kind permission of Dr. Mark Pelling; Plate 10.1 reproduced by kind permission of Liz Light.

Whilst every effort has been made to trace the owners of copyright material, in a few cases this has proved impossible and we take this opportunity to offer our apologies to any copyright holders whose rights we may have unwittingly infringed.

CHAPTER 1

Introduction

Katie Willis and Cathy McIlwaine

The region

Current debates in both development theory and development practice are often framed in thematic terms, rather than focusing on certain sectors or particular countries. While clearly aware of the great importance of local specificity, in this book we have followed the lead of the first volume in the DARG Regional Development Series (Simon and Närman 1999). Our aim is to consider the future development challenges and review past changes as they have been played out in the region that we have termed 'Middle America'. Many of the economic, political and social processes described are, or have been present in other parts of the 'developing world', but it is the interplay between the 'global', the 'regional' and the 'local' that results in the specific outcomes described in this text.

In approaching 'development', it is sometimes necessary to challenge pre-existing forms of categorisation to help both theory construction and policy implementation. In opting to focus on the nations of Central America (Belize, Costa Rica, El Salvador, Guatemala, Honduras, Nicaragua, Panama), Mexico and the Caribbean (islands in the Caribbean Sea plus the mainland territories of Guyana, French Guiana and Suriname) as one regional grouping, we are digressing from the conventional approach to an examination of this part of the world. A more usual framework is to place Central America and Mexico in the category of 'Latin America', with the island and mainland states of the Caribbean in a different category. However, the decision to produce a volume on Mexico, Central America and the Caribbean comes from both a belief that these nations share certain characteristics and a recognition that this part of the world is often considered as a footnote in discussions regarding the larger and more economically developed nations of South America. While Venezuela and Colombia are clearly

Caribbean littoral states, they are not included in our definition of Middle America. Some authors (for example, Clarke 1991) include Venezuela in 'the Caribbean', but the post-independence development trajectories and regional identifications of both Venezuela and Colombia mean that they are more usefully considered as part of 'South America'.

Although we will argue that it is logical to consider the nations of Central America, Mexico and the Caribbean as one region, this does not mean that we are oblivious to the diversity within them. Economically, the area encompasses the poorest state in the western hemisphere (Haiti), as well as states that cannot really be considered 'Developing world' in economic terms, such as the Bahamas (see Table 1.1). If the definition of 'development' is broadened to encompass indicators of standards of living, there are similar extremes. The Human Development Index (HDI) is a composite measure developed by the United Nations Development Programme (UNDP) and includes a measure of economic development (gross domestic product – GDP – per capita) as well as measures of educational attainment (adult literacy rates and educational enrolment rates) and health (life expectancy at birth). Within Middle America, three nations (Barbados, the Bahamas, and Antigua and Barbuda) are classified as having high levels of human development, while Haiti is found in the low human development category (see Table 1.1). Levels of urbanisation and the sizes of capital cities are also particularly diverse (see Pelling, this volume) and there are great variations in size of population, surface area and population density (see Table 1.1).

Given this diversity, why should we consider these countries within one regional volume? The rest of this chapter will consider the range of economic, political and social factors that bind these diverse nations together. These include a common history of colonialism, albeit by a number of European powers, as well as a shared experience of being within the shadow of the US and the geopolitical and economic implications of this proximity. Economically, the countries of the region have been characterised by a reliance on export agriculture, but in recent years have developed more varied strategies to adapt to changing global economic trends and the imposition of structural adjustment policies. The development strategies adopted have also shared similarities, although there are, of course, notable exceptions such as the Castro regime in Cuba and the Sandinistas of 1980s Nicaragua. Civil war and political conflict also characterised a number of the region's states during the twentieth century, especially in the 1980s, with the concomitant issues of 'postwar reconstruction' and returns to 'democracy'. Finally, there are examples of regional 'integration' both in economic terms through various trade agreements and external aid packages, but also in relation to migration systems.

We have chosen the term 'Middle America' to refer to the nations of Mexico, Central America and the Caribbean (Figure 1.1). Klak (1999) also uses this term in his discussion of recent trends in Central America and the Caribbean, but excludes Mexico from this categorisation because of its much larger economy and its inclusion in the North American Free Trade Agreement (NAFTA) alongside the US and Canada. While recognising that Mexico differs from some regional states in many ways, there are

Table 1.1 Diversity within Middle America

Country	GNP per capita 1998 (US$)	HDI 1998 (ranking 1998)[c]	Population 1998	Surface area 1996 (1,000 km²)	Pop. density (people per km²)
Antigua & Barbuda[b]	8,300	0.833 (37)	67,000	0.4	152
Bahamas, The[b]	(9,361+)	0.844 (33)	294,000	13.9	29
Barbados[b]	7,890	0.858 (30)	266,000	0.4	618
Belize	2,610	0.777 (58)	236,000	23	10
Cayman Islands[b]	(9,361+)	n.a.	36,000	0.3	138
Costa Rica[a]	2,780	0.797 (48)	4 million	51	69
Cuba[b]	(761–3,030)	0.783 (56)	11.1 million	110.9	101
Dominica[b]	3,010	0.793 (51)	74,000	0.8	98
Dominican Republic[a]	1,770	0.729 (87)	8 million	49	171
El Salvador[a]	1,850	0.696 (104)	6 million	21	292
Grenada	3,170	0.785 (54)	96,000	0.3	283
Guadeloupe[b]	(3,031–9,360)	n.a.	431,000	1.7	255
Guatemala[a]	1,640	0.619 (120)	11 million	109	100
Guyana[b]	770	0.709 (96)	857,000	215.0	4
Haiti[a]	410	0.440 (150)	8 million	28	277
Honduras[a]	730	0.653 (113)	6 million	112	55
Jamaica[a]	1,680	0.735 (83)	3 million	11	238
Mexico[a]	3,970	0.784 (55)	96 million	1,958	50
Netherlands Antilles[b]	(9,361+)	n.a.	213,000	0.8	266
Nicaragua[a]	(760 or less)	0.631 (116)	5 million	130	40
Panama[a]	3,080	0.776 (59)	3 million	76	37
Puerto Rico[b]	(3,031–9,360)	n.a.	3,857,000	9.0	435
St Kitts & Nevis[b]	6,130	0.798 (47)	41,000	0.4	113
St Lucia[b]	3,410	0.728 (88)	160,000	0.6	263
St Vincent & the Grenadines[b]	2,420	0.738 (79)	113,000	0.4	290
Suriname[b]	1,660	0.766 (67)	413,000	163.3	3
Trinidad & Tobago[b]	4,430	0.793 (50)	1,317,000	5.1	257

Sources: [a] World Bank (2000: Table 1, 230–231) for all figures except HDI. Bracketed figures are estimates; [b] World Bank (2000: Table 1a, 272) for all figures except HDI. Bracketed figures are estimates; [c] UNDP (2000: Table 1, 157–160). Bracketed figures show the country's 1998 ranking, the highest being 1.
Note
The Human Development Index (HDI) is calculated by the UNDP to provide a broader indicator of 'development' than the usual GDP per capita measures. The HDI includes measures for life expectancy at birth, educational attainment (measured through adult literacy rates and education enrolment rates), and GDP per capita. Countries are ranked from 1 to 174 according to their HDI ranking. According to 1998 figures, Canada came top with an HDI figure of 0.935. Sierra Leone had the lowest HDI figure of 0.252.

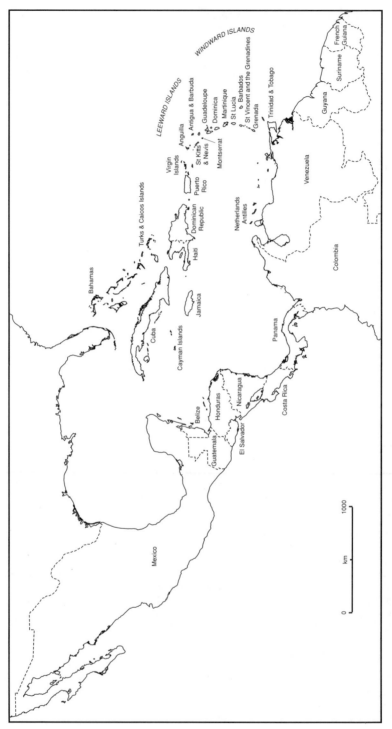

Figure 1.1 The countries of the Middle America region

numerous similarities due to shared characteristics both historically and in the present. These similarities will be drawn out throughout the book. Whilst we recognise that 'Middle America' may not be the most widely used term in the region itself, it does however provide a very useful short-hand for the countries of the region. An alternative term is 'the Caribbean Basin' (used by Grugel in her 1995 study of Central America and the Caribbean), but we have rejected this on the grounds that it was a relatively recent construction and too closely tied to US foreign policy formulations (see Klak 1995: 298).

Globalisation

The concept of 'globalisation' has framed much of the discussion about 'development' strategies and the role of the nation-state at the start of the twenty-first century (see for example, Hettne 1995). Globalisation also provides a context for much of the discussion in the later chapters, so needs to be examined in this introductory section.

It is important to recognise that, rather than being a coherent and homogeneous process, globalisation refers to a series of diverse processes (Held *et al.* 1999). All include the basic concept of the world becoming a more interconnected place, but the processes involved may be economic, social, political, cultural or environmental. The fact that globalisation is more than an economic concept is a key one. In economic terms, globalisation usually refers to the ways in which countries and regions are drawn more tightly into the world trading system, particularly as trade is liberalised and the role of multi- and transnational companies expands (Dicken 1998). However, this may be linked to changes in political processes; for example, the challenges to the nation-state due to the power of international corporations and multilateral organisations such as the World Bank or United Nations (UN). 'Cultural globalisation' is sometimes referred to as 'Americanisation' or 'Westernisation', with the expansion of consumption-oriented societies, and the global reach of particular consumer goods such as Nike sportswear and Coca-Cola. However, such concepts fail to recognise the resistances to such forms of 'cultural imperialism', processes of hybridisation and the ways in which 'non-western' cultural practices and traditions are also being transported due to globalisation (see for example, Tomlinson 1999). The rising popularity of 'world music' in the 'North' would be a good example of this, although as Klak (1999: 110) states, '[I]t is a stretch, however, to compare the northward influence of such artifacts as reggae music, Mayan handicrafts or merengue music with the multitude of cultural impacts in the opposite direction'.

This chapter, and many of the later chapters in this book, will discuss the myriad of ways in which the Middle America region has been incorporated into global economic, political and cultural processes since the beginning of European colonialism. This long-standing participation in extra-regional activities is used by some (so-called 'globalisation sceptics') to argue that there is nothing new about globalisation. However, other researchers point out that while these links at a global level do indeed have a long history, the interconnectivity of the globe, the speed of exchanges and the diversity of flows mean there is certainly something new about these processes at the end of the twentieth and the beginning of the twenty-first centuries (UNDP 1999). While this book does not set out to demonstrate the validity of the claims for globalisation in the context of Middle America, many of the contributors highlight the importance of new trends at a global level. It is crucial to recognise that while particular forms of globalisation may be perceived as beneficial in one location, or at one level, the processes involved often have adverse consequences for particular groups of people. Policies need to be implemented to harness the possibilities 'globalisation' has to offer (DFID 2000). In addition, while globalisation is often portrayed as an unstoppable juggernaut, many researchers have highlighted the ways in which the 'global' is constructed to meet 'local' needs (for example, Kelly 2000) and the need to step away from the rhetoric of globalisation to consider 'local world views and development strategies or ideologies that rely rather less on external determinants' (Simon 1999: 25; see also Mittelman and Othman 2000).

Colonial history

All the territories within the Middle America region have experienced colonial government by a European power at some point in their history. While experiences of colonialism vary, there are common factors and outcomes.

The majority of the region's lands were originally colonised by the Spanish during the sixteenth and seventeenth centuries. Following the arrival of Columbus in 1492, Spanish influence spread throughout the island Caribbean and on to the mainland, expanding until most of what is now Central and South America was under Spanish control. Spanish power also extended northwards through what is now Mexico and the southern states of the US (including Arizona, California, Nevada, New Mexico and Texas). This expansion of the Spanish empire was associated with the large-scale decimation of the indigenous populations (see Howard, this volume), and the establishment or expansion of urban centres (see Pelling, this volume). In addition, pre-existing mining and agricultural activities throughout the region were developed. This included silver mining in the Valley of Mexico and cacao production in the Pacific coastal regions of southern Mexico, Guatemala and El Salvador (Weaver 1994). Other export goods included indigo (especially from El Salvador), tobacco and gold. While Hispaniola (now Haiti and the Dominican Republic), Cuba and other Caribbean islands were developed economically by the Spanish, particularly through sugar and banana cultivation, they became much more important in the 1830s as military posts to protect the mainland colonies (Grugel 1995).

In South America, Spain's main challenge to colonial authority came from the Portuguese, but within Middle America the other main European influences were English, French and Dutch. While these European powers clearly had designs on the potential wealth of the region, many of the day-to-day threats to Spanish supremacy were from piracy. As Spanish sea-power declined, its European competitors were able to settle in the

Caribbean and seize Spanish-owned territories; for example, Jamaica became British territory in 1653. While the Spanish retained political control over Mexico and Central America, their economic influence was waning. For example, as the Industrial Revolution took hold, and in response to the relaxation of Spanish trade restrictions, English-produced cotton textiles were imported to Guatemala through Belize in the mid-eighteenth century. This undermined local textile production, a process which became all too familiar in the years to come (Weaver 1994).

All European colonial administrations permitted slavery in various forms. Under the Spanish *conquistadors* (colonisers), large numbers of indigenous people in Central America were enslaved, often being transported to Caribbean islands and Panama, as well as Peru (Weaver 1994). However, mass transportation of millions of West Africans to the region overshadowed these practices. An estimated 5.3 million slaves were imported into the Caribbean and the Spanish-dominated mainland between 1451 and 1870 (Curtin 1969). Following the emancipation of slaves in the British territories in 1833, indentured labour contributed to the ethnic diversity of the region. Many emancipated slaves migrated to the Caribbean coasts of Central America until the beginning of the twentieth century, especially to Nicaragua, Honduras, Costa Rica and Panama, to work on banana plantations. As well as Afro-Caribbeans, indentured labour from South Asia and China was also common, especially at the turn of the century (see Howard, this volume). Furthermore, it is sobering to remember that slavery remained legal in Puerto Rico until the 1870s, and in Cuba until the 1880s (Blouet 1997).

Independence for the mainland Spanish territories came in the early nineteenth century, following the wars of independence between the *criollos* (American-born Spaniards) and the Spanish authorities. From 1821 until 1824 Nicaragua, Guatemala, El Salvador, Costa Rica and Honduras were annexed to Mexico, and then formed a federation, the United Provinces of Central America. However, this regional alliance collapsed in 1838 as a result of internal policy disagreements.

The rest of the region had to wait until many decades later to achieve independence. The Dominican Republic gained independence in 1865 and the Spanish–American War of 1898 led to independence of a sort for Cuba and Puerto Rico, though Spanish external control was replaced by American influence, in both direct political terms and indirectly through economic relations.

For the British colonies, independence was a post-Second World War phenomenon, associated with the struggles for independence in other parts of the British Empire, most notably in sub-Saharan Africa and South Asia. Due to concerns over the small size of the Caribbean economies, 'independence' was first achieved through the Federation of the British West Indies (Jamaica, Trinidad and Tobago, Barbados, the Windward Islands and Leeward Islands), established in 1958. Jamaica seceded in 1961, Trinidad a year later, and the remaining islands gained their independence throughout the 1970s and 1980s (Grugel 1995). On the mainland, Guyana achieved independence in 1966, but Belize (previously British Honduras) gained full independence only in 1981.

While most territories have now achieved their independence, a number remain tied to another state in formal political terms. There are a number of dependent territories, including the French Antilles of Martinique and Guadeloupe, the Dutch Antilles of Aruba, Bonaire and Curaçao, and the British territories of Anguilla, Bermuda, the British Virgin Islands, Caymen Islands, Montserrat, and the Turks and Caicos Islands. Puerto Rico has internal self-government, but since 1917 Puerto Ricans have been classified as US citizens, with the US responsible for foreign affairs, defence and so on (Blouet 1997).

Development strategies: from import substitution to neoliberalism

At the time of independence, regardless of when it took place, all the region's economies were limited in size, vulnerable to external pressures and overly dependent on primary production. Grugel (1995) argues that these characteristics form a 'Caribbean model of development', and while there are some exceptions, most notably the size of Mexico's economy, there are many similarities in terms of development strategies adopted by the post-independence governments of the region.

The contributors to Simon and Närman's (1999) volume highlight the differing 'Developing World' approaches to development since the Second World War. From this it is evident that externally imposed theories and practices have been highly influential in many Southern countries. However, as Kay (1991) stresses, in Latin America, 'home-grown' theory played an important role in the construction of development strategies. The work of the Economic Commission on Latin America (ECLA), later expanded to ECLAC with the addition of the Caribbean, was key within this, with the work of Raúl Prebisch particularly prominent. From the 1940s onwards, ECLA adopted a structuralist approach to development in the region. Because of inequalities in the global trading system, the Commission argued that the region's countries would benefit from protectionism and a process of import substitution. Behind high tariff barriers, industrial and agricultural sectors could diversify without fear of foreign competition. Key elements within this process were interventions by national governments, and co-operation between countries which would reduce the region's dependence on external trade and assistance (see Kay 1989 for an overview).

Many Middle American countries followed these strategies (see Willis, this volume), but both internal and external factors have resulted in a shift away from such internally oriented policies to an approach that embraces the global market. For much of the post-Second World War period, the import substitution strategies were funded by foreign borrowing. Given the relatively low levels of international interest rates, and the seeming success of economic development policies, high levels of borrowing were regarded as sensible and relatively low risk.

However, in the late 1970s it became increasingly apparent that this development approach had limitations. Firstly, the import substitution policies adopted by many governments were no longer resulting in steady economic growth (see Willis, this volume). In addition, the region's links to the global economic system meant that it was vulnerable to adverse changes in external economic conditions, such as the rise in oil prices in 1973 and 1979, increases in global interest rates, and downturns in global trade. Under such conditions, the fall in export earnings and the rise in import bills and interest rates meant that many national governments could no longer meet the conditions for repaying their international borrowing (Roddick 1988).

The starting point of the resulting 'debt crisis' is usually regarded as 1982 when the Mexican Government defaulted on its loan repayment. However, commercial lending to many of the region's governments had been reduced for a number of years. For example, by 1979 Costa Rica's access to commercial finance was extremely limited (Sheahan 1987). In need of external finance, governments were compelled to adopt structural adjustment programmes (SAPs) approved by the International Monetary Fund (IMF). Successful adoption of these programmes would enable national governments to continue to borrow on international financial markets and to receive assistance from the IMF and World Bank. The underlying neoliberal ideology behind adjustment policies meant that governments were encouraged to open domestic markets to foreign trade and investment, devalue the currency (so the price of imports rises and the cost of exports declines) and reduce state expenditure (see McIlwaine, this volume; Willis, this volume).

This access to continued external finance and the restructuring of many of the region's economies eventually resulted in reduced debt burdens and positive growth rates (Inter-American Development Bank 1997). However, these macro-economic improvements were often at the expense of increasing income inequality and exacerbation of poverty (see McIlwaine, this volume). In addition, the ability to meet debt repayments does not mean that debt is no longer an issue for the region's governments. The debt burden is still significant for most countries (see Table 1.2), and although figures may be less severe than for many sub-Saharan African countries, the need to use public funds for debt payments results in less money being available for service provision (Esquivel *et al.* 2001). Nicaragua's debt burden stands out, with the amount of money owed to external sources representing over twice the GNP figure for 1997 and the debt service payments exceeding a quarter of export earnings in 1998. It is, therefore, unsurprising that Nicaragua has been included in the Heavily Indebted Poor Countries (HIPC) Initiative, a programme agreed by Northern governments and multilateral agencies to provide debt relief for the world's most-indebted nations. McIlwaine (this volume) discusses Nicaragua's involvement in this initiative. The UK Government's Department for International Development (DFID) has also made direct attempts to reduce debt burdens in the region, through the Commonwealth Debt Initiative launched in September 1997 (DFID 2001).

Neoliberal policies to reduce the involvement of the state in national economies and social services have also been reflected in changing forms of social and political organisation. With the decentralisation and privatisation of service provision, non-governmental organisations (NGOs)

Table 1.2 External debt

Country	External debt (US$ millions)		External debt as % of GNP		Total debt service as % of exports of goods and services	
	1986[a]	1997[b]	1986[a]	1997[b]	1986[a]	1998[c]
Costa Rica	3,889	3,548	97.8	34	28.9	7.6
Dominican Republic	2,756	4,239	55.5	27	21.7	4.2
El Salvador	1,547	3,282	40.2	25	20.8	10.4
Guatemala	2,306	4,086	31.7	21	24.3	9.8
Guyana	n.a	n.a	n.a	n.a	n.a	19.5
Haiti	585	1,057	27.4	21	6.0	8.2
Honduras	2,467	4,698	72.4	86	22.0	18.7
Jamaica	3,057	3,913	147.5	90	32.7	12.8
Mexico	91,062	149,690	76.1	37	51.5	5.0
Nicaragua	5,343	5,677	198.2	244	12.9	25.5
St Vincent and the Grenadines	n.a	n.a	n.a	n.a	n.a	13.7
Trinidad and Tobago	1,154	n.a	24.0	n.a	13.2	10.2

Sources: [a] World Bank (1988: Table 18, 256–257); [b] World Bank (2000: Table 21, 270–271); [c] UNDP (2000: Table 18, 219–222).

have often filled the void. This growth of NGOs is also heralded as a more democratic channel for development provision, as it allows for greater participation at the grassroots (see Bradshaw *et al.*, this volume). This move towards forms of 'bottom-up' development has also been reflected in a greater consideration of gender dimensions to development policies and the importance of sustainability (see Barton and Pelling, this volume).

In the context of Central America, Robinson (1998) discusses how the shift away from import substitution approaches to neoliberal development strategies is associated with a move from the 'national' to the 'transnational' as part of globalisation. As well as the increased participation of transnational corporations (TNCs) in national economics, there are other forms of transnational practices. These include changes in the state and political system with the increasing involvement of international financial institutions (such as the IMF and World Bank), and supranational organisations within the region (see below). In addition, social groupings are transformed by increasing external links; there is a development and an extension of a 'transnational elite', as well as the elaboration of 'transnational communities' created through outmigration (see below).

Any discussion of Middle American post-Second World War development strategies needs to highlight the anomaly that is Cuba after the 1959 revolution. The distinctiveness of the Castro regime in Cuba is not the existence of a socialist alternative to development in the region, as a number of Middle American governments have sought to follow such a path (see Willis, this volume), but its lengthy duration. The Sandinista regime lasted from 1979 until 1990 (see below), while Michael Manley's socialist experiment in Jamaica in the 1970s and Maurice Bishop's 'People's Revolutionary Government' in Grenada were even more short-lived. While Cuba has attempted to carve out an alternative development path, it is clear that in many ways recent changes have mirrored those in other parts of the region (see Willis, this volume).

Political conflict and violence

Broadly speaking, Middle America in the 1960s was economically and politically stable – a reflection of the buoyant regional and global economy and the generally democratic shifts to independence for the British colonies in the region. This stability faltered in the 1970s, not only with the oil crises of 1972 and 1979, but also with the emergence of civil wars, especially in Central America. Throughout this region, and especially in Nicaragua, Guatemala and El Salvador, the roots of civil war lay in deep-seated inequalities revolving around the historical concentration of political power, and especially land, in the hands of small oligarchies, often bolstered by US interests (see Grugel 1995). Throughout Central America, these political conflicts were waged between conservative states keen to maintain the status quo, and pro-Marxist revolutionaries whose aim was to challenge these inequalities and create more just societies. The presence of Castro's successful Cuban regime so close to home provided inspiration for many of these mainland left-wing organisations.

In Nicaragua, political conflict took the form of a revolution as the *Frente Sandinista de Liberación Nacional* (Sandinista National Liberation Front – FSLN) overthrew the dictatorial Somoza regime in 1979. In El Salvador and Guatemala, political conflict took the form of civil wars, with the popular classes waging guerrilla campaigns against the military and landowners, mainly in the light of severe and deteriorating socio-economic conditions. Starting in 1981, the civil war in El Salvador was waged between the *Frente Farabundo Martí para la Liberación Nacional* (Farabundo Martí Front for National Liberation – FMLN) and the repressive, authoritarian state. In Guatemala, the civil war dated back to the 1960s, and lasted for 36 years, although the conflict between the *Unidad Revolucionaria Nacional Guatemalteca* (Guatemalan National Revolutionary Unity – URNG) and the military was most active during the late 1970s and early 1980s. In Guatemala, the war evolved into a bloody and barbaric ethnic conflict, as the country's large indigenous population were targeted as revolutionaries. All these civil wars received military backing from the US, which financed the state in the cases of El Salvador and Guatemala, and the Contras (anti-Sandinista rebels) in Nicaragua (see Dunkerley 1994 for reviews of this period in Central America, and below on the role of the US).

A series of peace initiatives, instigated by the former President of Costa Rica, Oscar Arias, resulted in peace accords in both El Salvador (1992) and Guatemala (1996). In Nicaragua, the FSLN lost power to Violeta Chamorro and the *Partido Liberal Constitutionalista* (Liberal Constitutional Party – PLC) in a democratic election in 1990 (see Ardón 1999). Thus the 1990s became the decade of postwar reconstruction as these countries struggled to come to terms with their violent pasts and to rebuild their nations. The Zapatista uprising in Chiapas, Mexico, in 1994 is obviously a contradiction of this broad pattern, although the roots of the conflict, reflecting centuries of exclusion and inequalities, especially on the part of the indigenous population, have many similarities with the conflicts in other Central American countries in the 1980s (see Howard, this volume).

Reconstruction in Central America has taken a variety of forms. As noted above, most previously war-torn countries have reoriented their economies along neoliberal lines with concomitant implementation of SAPs (see Boyce 1995 on El Salvador). Widespread infrastructure projects such as road-building and increased access to education and healthcare provision have also been implemented throughout the affected countries. Civil society has played an important role in these reconstruction efforts as donor funding in particular is often channelled through NGOs (see also Bradshaw *et al.*, this volume).

The strengthening of civil society is also seen as an important component of societies characterised by widespread mistrust (McIlwaine 1998; Pearce 1998). In turn, with far-reaching psychosocial damage as a result of brutal wars (see Lira 1998 on Guatemala), efforts have been made to address psychological damage. In Guatemala, this has taken the form of the Recovery of Historical Memory Project (REHMI) by the Human Rights Office of the Archdiocese of Guatemala, involving the collection and publication of testimonies of survivors of the civil war (ODHAG 1999), providing a form of national catharsis for the country. Finally, reconstruction also involves dealing with returned refugees, many from camps in Mexico and Honduras, who initially fled the war-torn countries in the 1980s. With most refugees having returned to their own countries, challenges of reintegration remain (see Stein 1997 and below).

While this reconstruction has presented many challenges, the most serious has been the eradication or diminution of violence. Although peace was established and political violence reduced significantly, widespread social and economic violence has continued to affect Central America. Indeed, Pearce (1998) points out that murder rates in El Salvador were higher in the postwar 1990s than during the years of the civil war, with 9,000 murders per 100,000 people in 1994 compared with 6,000 during the conflict. Similar patterns have been noted elsewhere (see Box 1.1 on Guatemala). Guatemala is reportedly the most violent country in Central America (Moser and McIlwaine 2001), while Jamaica is probably the most violent country in Middle America as a whole, although it has not experienced political conflict (see Moser and Holland 1997). As well as murder rates, other types of violence and crime have been growing in the region. These include robberies, delinquency, intra-family and gender-based violence. Increasing drug production and distribution, as well as consumption, have become serious concerns, with signs that the problem will only worsen in the future (see Moser and McIlwaine 2001 on consumption in Guatemala; see below on production). In addition, the phenomenon of gangs – often referred to as *maras* in El Salvador and Guatemala – is a growing concern. Modelled on their US counterparts, especially the Los Angeles gang culture,

Box 1.1 Violence and crime in post-conflict Guatemala

Violence and crime in Guatemala remain serious problems despite the signing of peace accords in 1996. In a recent perception-based study conducted with 1,860 people in nine low-income urban communities throughout the country, violence-related problems emerged as the most pressing issues affecting poor communities (representing almost half of all the problems cited – 48 per cent). When people discussed the types of violence and crime affecting their communities, they identified those related to social violence, such as street fights, violence against women both inside and outside the home, and especially rapes in the street, as the most common (50 per cent of all types of violence), followed by economic violence, such as robbery, delinquency, as well as drug-related violence that was often economic in nature. Drug consumption rather than production or distribution was the main problem at the community level, involving mainly marijuana and crack, as well as glue and paint-thinner for those who could not afford the former. Overall, only 3 per cent of all violence identified was political, mainly pertaining to human rights abuse, especially in indigenous communities, as well as the phenomenon of lynchings – where groups of people in the community take the law into their own hands, and chase and kill (often by setting fire to them) those they suspect of committing acts of crime and violence. In general, the types of violence were extremely diverse, with focus groups listing on average 41 different types; in one community in Huehuetenango in the Central Highlands, 70 different types of violence were identified.

One of the single most commonly cited types of violence-related problems was youth gangs – known locally as *maras*. In Guatemala City in particular, these gangs often terrorised people in communities, stealing, mugging, taking drugs, getting drunk and behaving like delinquents. These *maras* have adopted the cultural traits of the US gangs from Los Angeles, with some gang members being recent deportees from the US. They use the street names of LA, such as Mara 18, named after 18th Street (and also a name commonly used by Salvadoran *maras*), as well as sporting the same tattoos and clothing (leather and bandanas). While the *maras* existed throughout the country, they were most violent and dangerous in Guatemala City, generating widespread fear in communities. In smaller towns and cities, the *maras* were not as ferocious, and often served as a pastime for unemployed youth, or as a rite of passage for young men (the gangs everywhere were male-dominated).

Source: Moser and McIlwaine (2001).

these Central American versions are often perceived as one of the most serious crime and violence issues in the region (see Box 1.1). Indeed, overall it should be emphasised that violence and crime present major development problems throughout Middle and Latin America, with potentially debilitating long-term economic and social consequences (Ayres 1998).

Within the islands and enclaves of the Caribbean, while violence and crime are a part of everyday reality for many, the civil war and authoritarianism found on the mainland have been much less prevalent in the post-independence period. This is a reflection of variations in social structures and colonial legacies. Within Central America and Mexico, the immediate post-independence period was characterised by the exacerbation of existing inequalities, based largely on land-holding. This pattern of social organisation was supported by the political system, and an authoritarian approach was adopted to prevent or control unrest (Torres-Rivas 1989). When this system of control broke down, political violence escalated. Within much of the English-speaking Caribbean, inequalities were less extreme at independence, and the legacy of a two-party parliamentary democracy and the union system helped channel discontent through formal structures (see Payne 1988 regarding Jamaica).

Regional co-operation

Given the small size of most of the region's nations, in both economic and physical terms, and the relative lack of power they hold on the international stage, forms of regional co-operation have been promoted since independence. While most of these have aimed to encourage economic development (see Willis, this volume), some have sought to strengthen the 'voice' of the region and develop common non-economic policies.

In the early days after independence there were forms of co-operation on both the mainland and within the island Caribbean. The United Provinces of Central America and the West Indies Federation were both established as alliances to assist the newly independent nations, albeit over a hundred years apart. As outlined above, both failed because of internal disagreements.

These unsuccessful attempts have not prevented further regional alliances, but all have had to deal with the need to balance national priorities with regional goals. Many authors (see for example, Bull 1999; Hettne *et al.* 1999; Montecinos 1996; Watson 1996) stress the importance of regionalism within a globalising world. They compare the inward-looking 'old regionalism' with the 'new regionalism' of the late twentieth and early twenty-first centuries which is outward-looking and is used as a platform for greater global integration. Within Central America the peace process of the 1980s provided an impetus for greater regional co-operation. The end to civil war in the region provided the stability necessary for greater integration, while the process of regionally based peace negotiations had developed lines of communication between the national governments (Bull 1999: 961). While the Central American Common Market (CACM) set up in the 1960s had largely been a forum for economic co-operation (see Willis, this volume), the 1990s have seen an increase in non-economic alliances on the isthmus. Bull (1999) outlines a series of agreements between the Central American nations during the 1990s, including the 1991 Protocol of Tegucigalpa that laid the groundwork for the System of Central American Integration (SICA) and the 1995 Treaty of Democratic Security. However, despite these good intentions, attempts at regional co-operation have had limited success, particularly because of the orientation of Central American nations towards other nations (especially the US) rather than their neighbours.

Since the break-up of the West Indies Federation, there have been other attempts at international co-operation within the island Caribbean. These have mainly focused on inter-regional trade arrangements such as CARIFTA (Caribbean Free Trade Area) and CARICOM (Caribbean Common Market) (see Willis, this volume), and have also played an important role in negotiating with external organisations such as the EU regarding preferential access for Caribbean agricultural products (see Thorpe and Bennett, this volume). However, there have also been regional groupings focusing on foreign policy, education etc. The University of the West Indies represents one attempt at fostering co-operation across islands, as does the continued fielding of West Indies cricket teams (Beckles 1999) and the survival of British West Indies Airlines (BWIA).

While the mainland and island nations have tended to have different forms of regional co-operation, largely as a result of their differing colonial histories, there have been recent moves towards a more cohesive Middle American grouping. In 1995 the first summit of the Association of Caribbean States (ACS) was held. The ACS includes the Central American nations, the Caribbean members of CARICOM and the Dominican Republic, in addition to Mexico, Colombia, Panama and Venezuela. Economic considerations have been an important part of this grouping, but there are other areas of co-operation, with special committees on natural disasters, tourism, environmental protection and other areas of overlapping interests (ACS 2000; see also Marshall 1998 on a Commonwealth Caribbean perspective on NAFTA, highlighting potential conflicts).

Relations with the US

While the proximity of Mexico to the US is often referred to as both a blessing and a curse, the US has also featured strongly in the geopolitical and economic concerns of the Central American and Caribbean nations. Throughout much of the twentieth century, US concerns regarding security in the face of threats from 'communist activity' in the western hemisphere led to a range of economic assistance policies, as well as direct and covert intervention in the domestic politics of the region.

The US government's involvement in the activities in other parts of the western hemisphere dates back to the early part of the nineteenth century, when the 1823 Monroe Doctrine warned the European powers to refrain from interfering in the Americas, in return for a similar lack of US involvement in European affairs (McCall 1984). It was only at the end of the nineteenth century that the US really had the military and economic might to act on this declaration, using the Monroe Doctrine as a justification for declaring war on the Spanish.

This position as 'protector of the Americas' became even more established in 1904, when US President Theodore Roosevelt announced his corollary to the Monroe Doctrine. In this he declared that Latin America was now within the US's sphere of influence, and that the US had the right to ensure political stability throughout the region by intervening where necessary. As the Cold War heightened after the Second World War, US military involvement in Middle America became increasingly apparent. The US government was either directly or indirectly involved in a number of political conflicts in the region in response to popular revolutions, or the elections of socialist governments committed to redistribution. Examples of intervention include the overthrowing of the democratically elected government of Jacobo Arbenz in Guatemala in 1954 (Trudeau 1984); funding of the *Contras* in Nicaragua during the 1980s; the invasion of Grenada in 1983 (Thomas 1988: 237–250); and the diplomatic and economic isolation of Cuba since the 1959 revolution.

Middle American states are not only influenced by the US through political intervention; economic dependence is also of great importance for all the region's countries. Mexico has a particular relationship with the US and Canada due to its involvement in NAFTA, which came into force on 1 January 1994. Under NAFTA, barriers to the trading of goods and services between the three members states were reduced, although specific items are still excluded from free trade, and the free movement of people is not included. NAFTA has increased Mexico's economic ties with the US, but the economic dependence of other Middle American states on the US is equally striking (see Table 1.3).

This economic dependence means that the region's economies are vulnerable to changes in the US domestic economy, as well as the use of economic tools for political ends, such as the trade embargo imposed on Cuba after the revolution. The US government has also regarded economic development as another tool to encourage political stability in the region. The Alliance for Progress was launched under the Kennedy administration in the 1960s. This included financial and technical assistance for health and education improvements and land reform programmes, for example. As well as underpinning ethical reasons for providing this support, the US

Table 1.3   Trade reliance on the US

Country	Exports to (%)	Imports from (%)
Belize (1997)	45.5	51.5
Dominican Republic (1996 and 1995)	44.7	44.1
El Salvador (1998)	54.3	52.8
Guatemala (1997)	47.6	45.6
Honduras (1999)	35.4	47.1
Jamaica (1997)	33.3	47.3
Mexico (1999)	88.4	74.3
Nicaragua (1999)	37.7	34.5
Puerto Rico (1996/7)	88.5	62.3
Suriname (1997)	15.5	35.4
Trinidad and Tobago (1998)	36.9	44.7

Sources: EIU (1998a, 1998b, 1999a, 1999b, 2000a, 2000b, 2000c).

Box 1.2   The Caribbean Basin Initiative

In 1983, the Reagan administration in the US launched the Caribbean Basin Initiative (CBI) to assist the region's economies in escaping their debt problems. The initiative was a three-pronged strategy consisting of investment incentives, trade opportunities and official aid (US$350 million on top of a budgeted US$474 million). However, while the CBI appeared to focus on economic matters, there was a strong political element, with certain countries (Cuba, Grenada and Nicaragua) being excluded, and others (notably El Salvador, Costa Rica and Jamaica) being given disproportionate assistance, because of the nature of their political systems.

The initiative was largely geared towards the development of external trade, particularly with the US, so fitting in with the neoliberalism of structural adjustment. Policies within the CBI included the abolition of duty for imports from the region into the US, as well as tax concessions for US investors. However, as with later negotiations regarding NAFTA, some goods (notably sugar, petroleum products and textiles) were excluded from the duty-free status because of pressures from domestic producers in the US. Instead, there was some renegotiation of quotas to allow greater imports from the region to the US, largely at the expense of producers in South America for sugar and the Far East for textiles.

The impact of the CBI has been regarded as decidedly limited, partly due to the exclusion of a number of the region's states, but also because the aid package, while seeming generous, could make few inroads into debt burdens. In addition, as 87 per cent of the region's exports already entered the US duty-free, the expansion of duty-free status made little difference.

For particular regional economies, however, there was a noticeable impact, although not always a positive one. Jamaica was one of the most favoured nations in the CBI as a result of the neoliberal policies adopted by President Edward Seaga. These policies, in conjunction with the CBI, led to an undermining of domestic manufacturing as a result of cheaper imports and the establishment of foreign-owned businesses. This has been associated with greater levels of unemployment and poverty.

Sources: Coppin (1992); Feinberg and Newfarmer (1984); Grugel (1995); Freeman (2000); Robinson (1998).

felt that this assistance would reduce the likelihood of political unrest (Sheahan 1987). The Caribbean Basin Initiative (see Box 1.2) of the 1980s was introduced for similar reasons, although it was also used as a channel for encouraging the adoption of neoliberal policies and as a way of reducing migration to the US (see below).

The final major axis of US–Middle American relations is drug trafficking. While coverage of US drugs policy in the Americas often highlights the Colombian drug cartels, it is clear that drug production also takes places in Middle America and drugs may be transported through the region en route to the US. Grugel (1995) states that 60 to 80 per cent of cocaine entering the US comes through the Caribbean Basin. Drug production within the region is also a major point of contention between the region's governments and the US; for example, by the mid-1990s it was estimated that Mexico produced 80 per cent of the marijuana imported into the US, and 20 to 30 per cent of the heroin consumed there (Smith 1997). Statistics like this have led to inter-governmental co-operation to counter drug trafficking and production operations. The US Drug Enforcement Administration (DEA) has played a high-profile role as part of these activities, assisting regional anti-drugs organisations in their intelligence operations and training staff. In March 2000, for example, 'Operation Conquistador' took place, involving 26 Caribbean nations, as well as Central and South American countries and the US. This two-week operation involved high levels of surveillance resulting in over 2,000 arrests and drug seizures, as well as training and co-operation between the participating nations (DEA 2001). Despite the concerns of regional governments regarding the drugs trade, the intervention methods of the DEA have not always been welcomed.

Migration systems

Regional linkages are evident in international migration flows of various kinds; for example, trading systems between Caribbean islands are developed and sustained by circular migration. However, other forms of movement are less voluntary, such as the flight of thousands of refugees from Central America to southern Mexico during the 1980s. Economic inequalities lie behind much of the migration from the region to the US, often creating what have been termed 'transnational communities'.

Inter-island trade has developed as a result of the limited markets found in some Caribbean islands. While national-level agreements regarding economic co-operation assist intraregional trade, individual migrants have also developed and sustained trade-related migration systems. In Trinidad and Tobago, for instance, 'suitcase traders' travel between Miami and other Caribbean islands selling electronic goods, clothing and other items not available in the Caribbean itself (Lloyd Evans 1998).

As mentioned earlier, the civil unrest in Central America in the 1980s led to thousands of refugees fleeing Guatemala, El Salvador and Nicaragua to Mexico. There were an estimated 45,000 Guatemalan political refugees living in UN camps in the 1980s, and as many as 150,000 other Guatemalans who migrated to Mexico but were not formally recognised as refugees (Jonas 1996). Others reached the US as illegal immigrants; although a significant number have since returned, some (especially Salvadoran gang members) have been deported, but most have remained in the US (see below). Natural disasters have also contributed to cross-border migration in Middle America. For example, the volcanic eruption in Montserrat in 1995 led to the displacement of large numbers of people. The population of the island fell from 10,400 in 1995 to 3,200 in 1997, largely because of migration to Antigua and other neighbouring islands, as well as to the UK, as part of a UK-sponsored evacuation (EIU 1999a).

Migration from the region to more economically developed countries has a long history; for example, there was significant migration from Mexico to the US at the end of the nineteenth century, and there were substantial numbers of migrants moving from the British Caribbean to the UK in the 1950s and 1960s. Despite increasing controls on immigration, migration to the North, particularly the US, stands at unprecedented levels. This is a response to the demand for cheap, unskilled labour in both urban and rural sectors, and also the supply of migrants willing and able to work in such activities, seeking a better life (Sassen 1988).

These international migrants are developing what have been termed 'transnational communities', as rather than migrating and losing contact with the 'home' country or village, they now often continue to have regular and sustained relationships in economic, familial and political spheres at home, as well as developing relationships in the place of destination (Glick Schiller *et al.* 1992; Portes *et al.* 1999). These migrants may engage in transnational practices such as remittances (Kaimowitz 1990; Robinson 1998), economic activities, political organisation and childcare (Hondagneu-Sotelo and Avila 1997). In addition, sending countries may embrace their overseas nationals as they represent an important source of foreign exchange and possible economic investors (see Mahler 1998). National identities may, however, be usurped by particular ethnic identities, which in turn intersect with gender: Guatemalan Mayan migrants to Los Angeles have experienced discrimination both in Guatemala and the US, leading them to develop a stronger Mayan identity, rather than associating with fellow Guatemalans (Popkin 1999).

Structure of the book

Many of the themes outlined in this introduction are picked up again in later chapters. Given the recent focus on grassroots approaches to 'development', and the need to understand 'local' priorities and conceptions of 'development' (see Simon and Närman 1999), we have largely structured the book so that the reader moves from the microlevel of individuals and households, through community considerations, to a focus on national-level strategies in different development areas. In Chapters 2 and 3, Sylvia Chant and David Howard consider the regional constructions of gender and ethnicity, and how these particular social characteristics have been incorporated into development policies. These themes of gender and race are then integrated into most of the subsequent chapters. The next two chapters consider issues of poverty (McIlwaine) and livelihoods (McIlwaine *et al.*). In both chapters, the authors are keen to stress the local specificity of processes and solutions, recognising the interactions between larger-scale national and global processes, and individual and household characteristics. Chapters 6, 7 and 8 focus on national-level development strategies, highlighting how governments have implemented policies in the agricultural sector (Thorpe and Bennett) and the non-agricultural sectors (Willis) in response to changing external forces and particular development ideologies. Jonathan Barton (Chapter 8) stresses the environmental implications of such policies, and how transformations in global and national environmental discourses are having an impact in the region. Mark Pelling (Chapter 9) continues this focus on sustainability, examining urban environments, but by using examples from a range of countries and scales he exemplifies the multidimensional nature of sustainable urbanisation. In Chapter 10, Sarah Bradshaw, Brian Linneker and Rebeca Zúniga use NGO activity in the reconstruction process after hurricane 'Mitch', to demonstrate the different scales at which civil society can operate, from the local through the national and regional to the global. The book concludes with a summary of the main themes and an identification of future areas of concern and hope. As a whole it aims to provide the reader with an overview of the economic, political, social, cultural and environmental changes that have occurred in the region, as well as highlighting the challenges Middle American nations are still facing at the start of the twenty-first century.

Useful websites

www.acs-aec.org Association of Caribbean States. Information about the member states of this organisation, as well as details of ACS working parties and regional agreements.

www.caribisles.org Website publicising eight Eastern Caribbean states (Antigua and Barbuda, British Virgin Islands, Dominica, Grenada, Montserrat, St. Kitts and Nevis, St. Lucia, St. Vincent and the Grenadines). Useful information about each state, and the Organisation of East Caribbean States.

www.eclac.cl UN Economic Commission for Latin America and the Caribbean. Available in Spanish and English. Useful information about UN research and policies in the region.

www.lanic.utexas.edu Latin American Network Information Center, University of Texas. An excellent starting point for basic information about countries of the region. Includes non-Hispanic Caribbean as well. Organised by country and by themes.

www.lib.nmsu.edu/subject/bord/laguia/ Internet Resources for Latin America. Another excellent starting point for regional websites.

www.transcomm.ox.ac.uk Economic and Social Research Council Transnational Communities Programme site. Excellent source of material on transnational communities, including those in the Americas.

Further reading

Grugel, J. (1995) *Politics and Development in the Caribbean Basin: Central America and the Caribbean in the New World Order*. Macmillan, Basingstoke. A clear overview of the similarities and differences within the Middle America region.

Robinson, W.I. (1998) (Mal)development in Central America: globalization and social change, *Development and Change* 29: 467–497.
An excellent overview of the ways in which globalisation processes have affected Central America.

References

Ardón, P. (1999) *Post-war Reconstruction in Central America: Lessons from El Salvador, Guatemala, and Nicaragua.* Oxfam, Oxford.

Association of Caribbean States (ACS) (2000) Homepage of the Association of Caribbean States *www.acs-aec.org.*

Ayres, R.L. (1998) *Crime and Violence as Development Issues in Latin America and the Caribbean.* World Bank, Washington D.C.

Beckles, H. McD. (1999) Whose game is it anyway? West Indies cricket and post-colonial cultural globalism, in Skelton, T. and Allen, T. (eds) *Culture and Global Change.* Routledge, London: 252–259.

Blouet, O. (1997) The West Indian region, in Blouet, B.W. and Blouet, O.M. (eds) *Latin America and the Caribbean: A Systematic and Regional Survey* (3rd edn). John Wiley, Chichester: 272–323.

Boyce, J.K. (1995) Adjustment towards peace: an introduction, *World Development* **23**(12): 2067–2077.

Bull, B. (1999) 'New Regionalisms' in Central America, *Third World Quarterly* **20**(5): 957–970.

Bulmer-Thomas, V. (1987) *The Political Economy of Central America since 1920,* Cambridge University Press, Cambridge.

Clarke, C. (1991) Introduction: Caribbean decolonization – new states and old societies, in Clarke, C. (ed.) *Society and Politics in the Caribbean.* St Antony's/Macmillan, Basingstoke: 1–27.

Coppin, A. (1992) Trade and investment in the Caribbean Basin since the CBI, *Social and Economic Studies* **41**(1): 21–43.

Curtin, P. (1969) *The Atlantic Slave Trade: A Census.* University of Wisconsin Press, Madison.

Department for International Development (DFID) (2000) *Eliminating Poverty: Making Globalisation Work for the Poor.* HMSO, London.

—— (2001) Homepage for the Department for International Development (*www.dfid.gov.uk*)

Drug Enforcement Administration (DEA) (2001) DEA Homepage (*www.usdoj.gov/dea*)

Dicken, P. (1998) *Global Shift: Transforming the World Economy* (3rd edn). PCP, London.

Dunkerley, J. (1994) *The Pacification of Central America*. Verso, London.

Economist Intelligence Unit (EIU) (1998a) *Dominican Republic and Puerto Rico: Country Report 1998–1999*. EIU, London.

—— (1998b) *Jamaica, Belize and the Organisation of Eastern Caribbean States: Country Profile 1998–1999*. EIU, London.

—— (1999a) *Jamaica, Belize and the Organisation of Eastern Caribbean States: 2nd Quarter 1999*. EIU, London.

—— (1999b) *Trinidad and Tobago, Guyana and Suriname: Country Profile 1999–2000*. EIU, London.

—— (2000a) *El Salvador and Guatemala: Country Profile 1999–2000*. EIU, London.

—— (2000b) *Mexico: Country Profile 2000*. EIU, London.

—— (2000c) *Nicaragua and Honduras: Country Report 2000*. EIU, London.

Esquivel, G., Larraín, F. and **Sachs, J.** (2001) Central America's foreign debt burden and the HIPC initiative, *Bulletin of Latin American Research* **20**(1): 1–28.

Feinberg, R.E. and **Newfarmer, R.** (1984) The Caribbean Basin Initiative: bold plan or empty promise?, in Newfarmer, R. (ed.) *From Gunboats to Diplomacy: New U.S. Policies for Latin America*. The Johns Hopkins University Press, Baltimore and London: 210–227.

Freeman, C. (2000) *High Tech and High Heels in the Global Economy: Women, Work and Pink-Collar Identities in the Caribbean*. Duke University Press, Durham and London.

Glick Schiller, N., Basch, L. and **Blanc-Szanton, C.** (1992) Towards a definition of transnationalism: introductory remarks and research questions, in Glick Schiller, N., Basch, L., and Blanc-Szanton, C. (eds) *Transnationalism: a New Analytic Framework for Understanding Migration, Annals of the New York Academy of Science* 645: ix–xiv.

Grugel, J. (1995) *Politics and Development in the Caribbean Basin: Central America and the Caribbean in the New World Order*. Macmillan, Basingstoke.

Held, D., McGrew, A., Goldblatt, D. and **Perraton, J.** (1999) *Global Transformations: Politics, Economics and Culture*. Polity Press, Cambridge.

Hettne, B. (1995) *Development Theory and the Three Worlds* (2nd edn.). Longman, Harlow.

Hettne, B., Inotai, A. and **Sunkel, O.** (eds) (1999) *Globalism and the New Regionalism*. Macmillan, Basingstoke.

Hondagneu-Sotelo, P. and **Avila, E.** (1997) 'I'm here, but I'm there': the meanings of transnational motherhood, *Gender and Society* 11: 548–571.

Inter-American Development Bank (IDB) (1997) *Latin American After a Decade of Reforms*. IDB, Washington D.C.

Jonas, S. (1996) Transnational realities and anti-immigrant state policies: issues raised by the experiences of Central American immigrants and refugees in a trinational region, in Korzeniewicz, R.P. and Smith, W.C. (eds) *Latin America in the World-Economy*. Praeger, London: 117–132.

Kaimowitz, D. (1990) The 'political' economies of Central America, *Development and Change* 2: 637–655.

Kay, C. (1989) *Latin American Theories of Development and Underdevelopment*. Routledge. London.

—— (1991) Reflections on the Latin American contributions to development theory, *Development and Change* 22: 31–68.

Kelly, P. (2000) *Landscapes of Globalisation*. Routledge, London.

Klak, T. (1995) A framework for studying Caribbean industrial policy, *Economic Geography* **71**(2): 297–316.

—— (1999) Globalization, neoliberalism and economic change in Central America and the Caribbean, in Gwynne, R. and Kay, C. (eds) *Latin America Transformed: Globalization and Modernity*. Arnold, London: 98–126.

Lira, E. (1998) Guatemala: uncovering the past, recovering the future, in Eade, D. (ed.) *From Conflict to Peace in a Changing World*. Oxfam Working Paper, Oxfam, Oxford: 55–60.

Lloyd Evans, S. (1998) Gender, ethnicity and small business development in Trinidad: prospects for sustainable job creation, in McGregor, D., Barker, D. and Lloyd Evans, S. (eds) *Sustainability and Development in the Caribbean: Geographical Perspectives*. University of West Indies Press, Mona, Jamaica: 3–25.

Mahler, S. (1998) Theoretical and empirical contributions toward a research agenda for transnationalism, in Smith, M.P. and Guarnizo, L. (eds) *Transnationalism From Below*. Transaction Publishers, New Brunswick: 64–100.

Marshal, D.D. (1998) NAFTA/FTAA and the new articulations in the Americas: seizing structural opportunities, *Third World Quarterly* **19**(4): 673–700.

McCall, R. (1984) From Monroe to Reagan: an overview of US–Latin American relations, in Newfarmer, R. (ed.) *From Gunboats to Diplomacy: New U.S. Policies for Latin America*. The Johns Hopkins University Press, Baltimore and London: 15–34.

McIlwaine, C. (1998) Contesting civil society: reflections from El Salvador, *Third World Quarterly* **19**(4): 651–672.

Mittelman, J. and **Othman, N.** (eds) (2000) Capturing globalization. Special issue of *Third World Quarterly* **21**(6).

Montecinos, V. (1996) Ceremonial regionalism, institutions and integration in the Americas, *Studies in Comparative International Development* **31**(2): 110–123.

Moser, C. and **Holland, J.** (1997) *Urban Poverty and Violence in Jamaica.* World Bank, Washington D.C.

Moser, C. and **McIlwaine, C.** (2001) *Violence in a Post-Conflict Context: Urban Poor Perceptions from Guatemala.* World Bank, Washington, D.C.

ODHAG (Human Rights Office of the Archdiocese of Guatemala) (1999) *Guatemala: Never Again!.* Catholic Institute of International Relations in association with the Latin America Bureau, London.

Payne, A. (1988) *Politics in Jamaica.* Hurst, London.

Pearce, J. (1998) From civil war to 'civil society': has the end of the Cold War brought peace to Central America?, *International Affairs* **74**(3): 587–615.

Popkin, E. (1999) Guatemalan Mayan migration to Los Angeles: Constructing transnational linkages in the context of the settlement process, *Ethnic and Racial Studies* **22**(2): 267–289.

Portes, A., Guarnizo, L. and **Landolt, P.** (1999) The study of transnationalism: pitfalls of an emergent research field, *Ethnic and Racial Studies* **22**(2): 217–237.

Robinson, W.I. (1998) (Mal)development in Central America, *Development and Change* 29: 467–497.

Roddick, J. (1988) *The Dance of the Millions: Latin America and the Debt Crisis.* Latin American Bureau, London.

Sassen, S. (1988) *The Mobility of Labour and Capital: a Study in International Investment and Labour Flow.* Cambridge University Press, Cambridge.

Sheahan, J. (1987) *Patterns of Development in Latin America: Poverty, Repression and Economic Strategy.* Princeton University Press, Princeton, NJ.

Simon, D. (1999) Development revisited: thinking about, practising and teaching development after the Cold War, in Simon, D. and Närman, A. (eds) *Development as Theory and Practice: Current Perspectives on Development and Development Co-Operation.* Longman, Harlow: 17–54.

Simon, D. and **Närman, A.** (eds) (1999) *Development as Theory and Practice: Current Perspectives on Development and Development Co-Operation.* Longman, Harlow.

Smith, P.H. (1997) Drug trafficking in Mexico, in Bosworth, B.P., Collins, S.M. and Lustig, N.C. (eds) *Coming Together? Mexico–US Relations.* Brookings Institute Press, Washington, D.C.: 125–154.

Stein, B. (1997) Reintegrating returning refugees in Central America, in Kumar, K. (ed.) *Rebuilding Societies after Civil War: Critical Roles for International Assistance.* Lynne Rienner, Boulder, CO, and London: 155–180.

Thomas, C. (1988) *The Poor and the Powerless: Economic Policy and Change in the Caribbean.* Latin American Bureau, London.

Tomlinson, J. (1999) *Globalization and Culture.* University of Chicago Press, Chicago.

Torres-Rivas, E. (1989) Authoritarian transition to democracy in Central America, in Flora, J. and Torres-Rivas, E. (eds) *Central America: Sociology of Developing Societies*. Macmillan, Basingstoke.

Trudeau, R. (1984) Guatemala: the long-term costs of short-term stability, in Newfarmer, R. (ed.) *From Gunboats to Diplomacy: New U.S. Policies for Latin America*. The Johns Hopkins University Press, Baltimore and London: 54–71.

UNDP (1999) *Human Development Report 1999*. Oxford University Press, Oxford.

—— (2000) *Human Development Report 2000*. Oxford University Press, Oxford.

Watson, H. (1996) Globalization, new regionalization, restructuring and NAFTA: implications for the signatories and the Caribbean, *Caribbean Studies* **29**(1): 5–48.

Weaver, F.S. (1994) *Inside the Volcano: the History and Political Economy of Central America*. Westview Press, Oxford.

World Bank (1988) *World Development Report 1998*. Oxford University Press, Oxford.

—— (2000) *Entering the 21ˢᵗ Century: World Development Report 1999/2000*. Oxford University Press, Oxford.

CHAPTER 2

Men, women and household diversity

Sylvia Chant

Introduction

This chapter focuses on households in Middle America, with particular emphasis on the role of gender in explaining contemporary diversity in household composition, headship and organisation. Although most of the statistical material pertains to the region as a whole, the discussion concentrates primarily on Mexico, Central America and the Hispanic Caribbean, with many of the case study examples drawn from recent research by the author on gender, masculinities, and 'the family' in northwest Costa Rica.[1]

The chapter first picks out key patterns and trends in household structure in different parts of Middle America, and identifies major economic, social and demographic factors bearing upon household form, organisation and dynamics. The next section explores the inter-relationships between households and gender. Particular reference is made to divisions of labour, power and resources, the construction of familial gender identities, and the manner in which recent changes in gender roles and relations are serving to erode in principle, if not always in practice, the idealised norm of a 'traditional' patriarchal family unit. The final section examines policy responses of state and civil society organisations to changing household realities, including social programmes and poverty alleviation initiatives.

[1] This research comprises the following projects: 'Men, Households and Poverty in Costa Rica: A Pilot Study' (1997), funded by the Nuffield Foundation (SOC/100[1554]), and ESRC (Award No. R000222205); 'Institutional Perspectives on Family Change in Costa Rica (1999), funded by the Central Research Fund, University of London and Department of Geography and Environment, LSE, and 'Youth, Gender and "Family Crisis" in Costa Rica' (1999) funded by the Nuffield Foundation (SGS/LB/0233). The author makes grateful acknowledgement for the financial support granted by these organisations, as well as to Carlos Borge who was field assistant on the 'Men, Households and Poverty' project, and Wagner Moreno who assisted in the project on 'Youth, Gender and "Family Crisis" in Costa Rica'.

Although some attention is given to class differences, the main focus of the discussion is low-income households, which make up the bulk of the population in the region.

Household diversity in Middle America

Household diversity is a long-standing feature of the social and demographic landscape of Middle America. Although most households in the region, past and present, have been family based, a distinction is often drawn between the predominantly female-headed and/or female-centred units of the Caribbean and Afro-Caribbean communities along Middle America's Atlantic coast (McIlwaine 1993; Safa 1999; Smith 1988), and the predominantly male-headed, patriarchal households of Central America and Mexico (Chant 1997; González de la Rocha 1999; Rudolf 1999). Yet on closer scrutiny there is often as much diversity within, as between, these sub-regions. As Ellis (1986: 7) points out for the Caribbean:

> Family forms . . . are very diverse. The European concept of the nuclear family, its ideology of patriarchy and male dominance, the women-centred matriarchal type and the extended family type, legacies from African and East Indian cultures respectively, all exist simultaneously within the region.

These observations are, in essence, equally applicable to Central America and Mexico. Here both Hispanic and indigenous cultural influences have intermeshed with differing economic, demographic and political circumstances to produce a vast array of household types, not to mention varied and often elaborate networks of exchange with kin (see Dore 1997; González de la Rocha 1994; Safa 1999; Willis 1993, 2000; also Boxes 2.1 and 2.2).

Beyond recognising diversity in household form per se, the contractual arrangements on which they are based may also vary. Although legal marriage is usually regarded as a social ideal (and in some countries, such as Mexico, is the norm in practice), in many instances partnerships between men and women have taken the form of consensual unions. In Cuba, for example, fewer than three in five couples marry legally (Lumsden 1996: 44).

Factors contributing to the lack of formal marriage in the case of Central America include indigenous antecedents, such as the non-practice of marriage among the Chorotegas of northwest Costa Rica and southern Nicaragua (Chant 1997: 170). The financial costs of marriage proceedings, especially in isolated rural communities lacking a church or permanent clerical presence, have also played a role (Rudolf 1999). Another factor is the perceived irrelevance of life-long contracts in situations where people's economic status is precarious on account, *inter alia*, of the seasonality of tropical agricultural production (Jayawardena 1960). For example, aside from the practices of their Chorotegan forebears, one reason for the historic instability of conjugal relationships in Guanacaste, Costa Rica, is that the province has traditionally offered little in the way of regular employment. One of the least urbanised areas in the country, Guanacaste

---

Box 2.1 Commonly occurring household
structures in contemporary Middle America

Household structure	Brief description
Nuclear household	Couple and their biological children
Blended/step-family household	Household in which one or both partners in a couple is not the biological parent of one or more co-resident children.
Female-headed household	Generic term for a household where the senior woman or household head lacks a co-resident male partner. Often, although not always, household head is a lone mother (see also Table 2.2)
Extended household	Household which in addition to one or both parents and children comprises other blood relatives or in-laws (*de facto* or *de jure*). May be male- or female-headed, laterally or vertically extended, and/or multi-generational
Nuclear-compound household	Arrangement where two or more related households share the same living space (for example, dwelling or land plot), but operate separate household budgets and daily reproduction (for example, cooking, eating)
Grandmother-headed household	Grandmother and her grandchildren, but without intermediate generation
Lone/single-person household	Woman or man living alone

Sources: Brydon and Chant (1989); Chant (1991, 1992).

---

was until recently dominated by cattle-ranching, sugar cultivation, and irrigated rice production. Sugar cultivation is the only labour-intensive activity, and then only on a seasonal basis, with significant recruitment of workers generally restricted to the harvest period between January and April. For low-income men, who comprise the bulk of workers in agriculture, and on whom the responsibility for household 'breadwinning' has conventionally fallen, this has implied migration to other parts of the country for spells of varying duration in order to support their families. Yet although *Guanacastecos* are commonly found working in the banana plantations on the Caribbean or southern Pacific coasts, or clearing pasture in the wetter zones of Puntarenas province, work is not always easy to come by (especially for more than a day at a time), leading to frustration and hardship for migrants and their families alike. The situation is aggravated by the fact that prolonged periods of separation place extra pressures on couples. It is not uncommon for men to cease sending remittances after time has elapsed, and sometimes not to return home at all. Although this is occasionally because they feel they cannot go back to

Box 2.2   Typology of female-headed households

Type	Brief description	Frequency among female-headed households in Middle America
Lone mother household	Mother with co-resident children	High
Female-headed extended household	Household comprising lone mother, children and other relatives	High
Lone female household	Woman living alone (usually elderly)	Low but increasing
Single sex/female-only household	Woman living with other women (female relatives or friends)	Low
Lesbian household	Woman living with female sexual partner	Low
Female-dominant/ predominant household	Household headed by woman, where although males may be present, they are only junior males with less power and authority than adult females	Variable – high in Caribbean, low in Central America and Mexico
Grandmother-headed household	Grandmother and her grandchildren, but without intermediate generation	Variable – moderate in Caribbean, low in Central America and Mexico
'Embedded' female-headed unit	Unit comprising a young mother and her children contained within larger household (usually that of parents). Sometimes referred to as 'female-headed sub-family'	Moderate

Source: Chant (1997: 10–26).

their families with nothing to show for their efforts, more usually it is because they have set up home with other women (Chant 2000).

In Nicaragua, too, where Lancaster (1992: xiv) describes the traditional family structure as 'both patriarchal and brittle', it is often the case that men initiate conjugal breakdown, commonly moving on to the households of women with whom they have already begun sexual relationships. A central factor in this pattern of 'informal polygamy' is the culture of *machismo* in the country (ibid.). Yet Vance (1987: 141–2) further notes that the instability of the Nicaraguan household is 'not merely a question of the irresponsibility of men or attributable to cultural attitudes', but is linked to the expansion of plantation agriculture during the Somoza regime, the mechanisation of farming production, the proletarianisation of the rural labour force, and the enforced migration of men in search of wages.

Consensual unions are not always linked with adversity, however; nor are they necessarily determined by men. In Nicaragua, for example, men often choose to get married to deter women from leaving. Women, on the other hand, may resist marriage because it gives them less leeway to bargain with their spouses (Lancaster 1992: 46). In much of the Caribbean, consensuality is again often the result of women's preferences not to live with men, but instead to opt for more flexible 'visiting unions' (Powell 1986). This is not just the case among the poor, but among middle-income and elite groups as well (Ellis 1986; Pulsipher 1993; Trotz 1996). While consensual unions can be as, if not more, enduring than marriages, the absence of legal barriers to splitting up means that they are often associated with temporary relationships between men and women (Chant 1992). In turn, serial conjugal partnerships give rise to people's homes being shared with a number of partners over the life course, such that large numbers of households contain step-children and half-siblings (Lumsden 1996: 121). These are sometimes referred to as 'blended' households, and can form part of larger multi-generational extended units (see Box 2.3; also Box 2.1).

As indicated in Box 2.3, extended households may be laterally as well as vertically extended, and comprise a complex range of close and more distantly related individuals. A recent study of *ejidatarias* (female members of agrarian co-operatives) in Mexico, for example, showed that in a sample of 487 households, as many as 131 were extended. This latter group, in turn, comprised over 30 different types of extended units, even if the majority were formed around a nexus of daughters or sons-in-law and grandchildren (Robles Berlanga *et al.* 2000: 51). Although extended households (as with other non-nuclear units such as female-headed households), are often viewed as being more common among low-income groups, Lomnitz and Pérez-Lizaur's (1991) study of Mexico indicates that the 'grand-family' or three-generation descent group is a distinctive feature of middle-income and elite sectors of society as well.

Female-headed households are another important household sub-group in Middle America, and are not just a late twentieth-century phenomenon (see Dore 1997 and Gudmundson 1996 on nineteenth-century Nicaragua and Costa Rica respectively). None the less, their growth has been fairly ubiquitous throughout the region in recent decades and is arguably one of the most prominent trends in household change in recent times (see below).

Where the extent of Middle American household diversity reaches something of an impasse, however, is when it comes to homosexual households. This is arguably due in part to the fact that this group have not to date been a major subject of research. In relation to homosexual men, for example, Alfredo Mirandé (1997: 132) suggests that: 'Perhaps the most glaring omission in research and writing on gender and masculinities is the absence either of research or writing on Latino gay men or of attempts to articulate a Latino male gay voice.' Moreover, what little has been written on homosexuality has tended to focus on the issue of identity and sexual practices, rather than household arrangements (ibid.; see also

Box 2.3   The evolution and characteristics of a typical low-income household in northwest Costa Rica

Maiela and Victor are presently in their early forties and live in a low-income settlement on the outskirts of Liberia, the capital of Guanacaste province in northwest Costa Rica. As of 1999, their household consisted of Maiela's daughter (Victor's stepdaughter) Ana Lorena (26 years), Ana Lorena's son José Gabriel (7 years), and her partner of five years Omar, who is not the biological father of the child. Maiela and Victor also have a daughter in common, Yorlenes (24 years), who until recently shared their house with her four-year-old son, Alan Emir, and former partner Melvin.

Like most couples in the settlement, Maiela and Victor are in a consensual union, rather than formally married. Somewhat less typically, however, their partnership is one that has stood the test of time. Victor is the only man Maiela has ever lived with, and they have been together 27 years, despite the origins of their relationship in less than auspicious circumstances.

Maiela was raised in a lone-parent household, but lost her mother at the age of 13. Although she continued to live with her elder siblings, lack of domestic vigilance and protection meant that Maiela's transition to adulthood came abruptly and traumatically. Following the death of her mother, a door-to-door pig seller who had been accustomed to visiting the house previously took advantage of Maiela's vulnerability and persuaded her to sleep with him. Maiela declared that while she was grateful for his attentions, at 14 years of age she was too young to know what she was doing and actually got pregnant on the first occasion. Two months later when she realised she was carrying a child and told the man in question, he promptly extricated himself from the *noviazgo* (courtship) and she never saw him again. Once her family learned about Maiela's condition, things became very difficult for her. One of her brothers, a priest, said that he had no wish to live in the same house as *'una puta'* ('harlot'), and that she had brought so much shame on the family he would probably have to cast off his habit and renounce his profession.

Fortunately for Maiela this unhappy situation was not to last. In a part-time cleaning job she took on at a local dance hall she came to know Victor. After telling him how difficult things were for her at home, he suggested they move in together. He would look after the child she was carrying and treat it as his own. Maiela jumped at the chance. Although there was talk of marriage, as there is still from time to time, this would require Victor to amend an erroneous birth certificate which would cost in the region of US$500. This has never been affordable however. Only recently did Victor manage to procure a stable, full-time job as a security guard. Prior to this he was reliant on casual work in agriculture and haulage, and Maiela has only worked sporadically as a domestic servant. As it is, Maiela feels that a free union is more or less the same as a marriage and perhaps even better because one can escape if need be. In the early stages of her relationship, this had crossed her mind, with Victor himself admitting to having being something of a *'vaguito'* ('bad boy') and *'mujeriego'* ('womaniser'), drinking, smoking and spending time with his friends *'en la calle'* (in the street). When Yorlenes was born, however, Victor became more responsible, and any thought that Maiela may have had of leaving was eclipsed by the notion that they should stick together for their daughter's sake. Interestingly, when Yorlenes was intending to marry her first partner, Melvin, she commissioned a marriage portrait of Victor and Maiela, as well as her own, which to this day hangs in their sitting area. Maiela is content within her union, and feels that the individual relationship a woman has with her spouse is more important than any law that has been passed for women's rights in Costa Rica in the last decade. She is also happy that her daughters have retained such close ties with her over the years.

Both daughters have certainly benefited from Maiela and Victor's 'open door' policy. Although both daughters left school and started work at an early age, as well as becoming mothers by their twentieth birthdays, they have both spent varying spells in their parents' home in adulthood, with partners and children. The younger of Maiela's daughters, Yorlenes, resided with Maiela and Victor between 1995 and 1997 when she first formed a relationship with Melvin, the father of Alan Emir. This was partly through lack of money, and partly because of parental resistance to her moving out. Maiela and Victor were concerned about Melvin when he first came on the scene, mainly because they could see he was a 'layabout', and spent too much time with his friends. On top of this, Yorlenes was suffering complications with her pregnancy and they did not trust Melvin to look after her properly. The birth of Alan Emir had a seemingly profound effect on Melvin who resolved to 'reform' his 'bad ways'. He even encouraged Yorlenes to move out to build their own house on a plot of land owned by her family a couple of streets away. Yet despite his intentions to start a new life, his good behaviour did not last. By the time Alan Emir was two-and-a-half, Melvin had started drinking again and seeing other women, forcing Yorlenes to terminate the relationship. Although Melvin is now living in the capital, San José, he phones continually and pleads with her to come back. However, Yorlenes found a new boyfriend, Minor, in 1999 with whom she now shares her new home. Minor is quite accepting of Alan Emir and they hope to give him a half-brother or -sister in the near future. When Yorlenes moved out of Maiela and Victor's house, Ana Lorena moved back in with her son and second partner. Although Ana Lorena had first left home at 14 to work in San José as a domestic servant, and actually has her own house in Liberia, because her settlement is recently established and still lacks basic services she prefers to stay with Maiela until such time as she is able to raise her son in a healthier environment. Maiela is perfectly happy about this as well as feeling that it is a right for Ana Lorena who sent money back home during her entire time in San José. For the foreseeable future, therefore, Maiela and Victor's household will continue to retain its 'blended/extended' character.

Source: Author's interviews (1989, 1992, 1994, 1996, 1999).

Buffington 1997; Lancaster 1997; Prieur 1996; Quiroga 1997). Accepting lack of visibility in research however, it would also appear that both male and female homosexual households are rare in practice in Middle America. In Cuba, for example, homophobia along with pressure on gays to be discreet means that very few live openly with one another (Lumsden 1996). This also applies to Costa Rica and is again undoubtedly due to anti-gay sentiment. According to research conducted under the auspices of the Central American Programme for Sustainability, Costa Rica's long-standing reputation for 'democratic values' sits in glaring contrast to the fact that its citizens are much less tolerant of minority groups than other Central Americans, with homosexuals ranking top of the list of 'most disliked groups', Nicaraguans second, and atheists third.[2] In turn, the role of the Roman Catholic Church in fomenting homophobia has been anything but subtle. In his Easter address to the nation in 1998, the Archbishop of San José, Monseñor Román Arrieta, described San José as having become an: *'antro de putrefacción'* ('den of iniquity', literally 'putrefaction'), rivalling major global cities for 'anti-social' vices such as 'the prostitution of minors, alcoholism, drug addiction, homosexuality and lesbianism'.[3]

Middle American households: complexities of continuity and change

The task of documenting, let alone evaluating, recent trends in Middle American household patterns is not easy, mainly because many contemporary as well as historical records are superficial and sketchy, with disparities between countries, as well as over time (Chant 1997: 149). Another factor is that the baselines for change – where they can be established – are often highly complex and differentiated. In addition, many different, and often countervailing, changes are occurring. As Radcliffe (1999: 200) notes for Latin America in general: 'Household forms have not changed in any easily discernible or unitary direction. In some areas family size has shrunk, and, in others, family members are joined by relatives, friends and other *allegados* (co-residents).' Moreover, diverse trends can occur within the same country or even town, on account of cross-cutting influences of class, ethnicity, gender, age and migrant status.

For example, although crisis and economic restructuring in the 1980s and 1990s saw increases in extended households in low-income communities in Mexican cities such as Guadalajara (González de la Rocha 1988) and Querétaro (Chant 1996), the proportion of extended households nationally dropped from 9.3 to 8 per cent between 1960 and 1990. This has been attributed partly to unfavourable conditions for multi-generational households in cities, and partly due to improving health among older age groups which may have encouraged elderly citizens to adopt a more independent lifestyle. Indeed, whereas 9 per cent of the population in the 65-year plus

[2.] Article by Mauricio Herrera Ulloa in *La Nación*, 15 August 1999, based on the book *El Sentir Democrático: Estudios Sobre la Cultura Política Centroamericana* by Florisabel Rodríguez.
[3.] Extract from a speech taken from a newspaper article 'Iglesia contra el pecado', by Emilio Mora, *La Nación*, 7 April 1998.

Table 2.1 Fertility and contraception in selected Middle American countries

Country	Total fertility rate*				Contraceptive prevalence rate (%) 1987–94**
	1970	1980	1994	2000	
Barbados	–	–	1.8	–	–
Costa Rica	4.9	3.7	2.9	–	75
Cuba	–	–	1.5	–	70
Dominican Republic	6.1	4.2	2.9	2.7	56
El Salvador	6.3	5.3	3.8	3.4	53
Grenada	–	–	–	–	54
Guyana	–	–	2.5	–	–
Guatemala	6.5	6.5	5.2	4.6	32
Haiti	–	5.2	4.8	–	18
Honduras	7.2	6.5	4.7	4.0	47
Jamaica	5.3	3.7	2.6	2.1	62
Mexico	6.5	4.5	3.2	2.6	53
Nicaragua	6.9	6.2	4.9	4.2	49
Panama	5.2	3.7	2.7	2.5	64
St Kitts & Nevis	–	–	–	–	41
St Lucia	–	–	–	–	47
St Vincent	–	–	–	–	58
Trinidad and Tobago	–	–	–	2.1	53

Sources: UNDP (1997: 194–5); World Bank (1995: 212–3); World Bank (1996: 198–9).

Notes
* The total fertility rate refers to the average number of children likely to be born to a woman if she survives until the end of her childbearing years and gives birth at each age in accordance with prevailing age-specific fertility rates
** Contraceptive prevalence refers to the proportion of women who are practising, or whose spouses are practising any form of contraception. This is generally measured for married women aged between 15 and 49 years
– = no data

age group in Mexico live alone, this is only 1 per cent across the population as a whole (López and Izazola 1995: 56). The increase in lone-person households looks set to continue with demographic ageing and, if gender differences in life expectancy persist, then it is also likely that more women will live alone than men; for example already as many as 27.5 per cent of female household heads in Costa Rica reside independently, compared with only 10.9 per cent of male heads, although this is partly a function of the fact that there are three times as many male-headed households as female (PEN 1998: 44).

Alongside the role of lone-person households in reducing household size in Middle America, other important factors include falling fertility. This is strongly associated with increased rates of contraceptive usage, often driven by state-sponsored family-planning programmes, dating as far back as the 1950s in Puerto Rico, and to the 1970s in Mexico (Côrrea and Reichmann 1994: 30; see also Table 2.1). In Puerto Rico, mainly poor,

Table 2.2 Women's average age at marriage and incidence of female household headship in selected Middle American countries

Country	Women's average age at marriage (years)		Proportion of households headed by women (%)	
	*1970**	*1990–2*	*1980**	*1990–2*
Barbados	–	–	43.9	–
Costa Rica	21.7	–	17.6	20.0
Cuba	19.5	–	28.2	–
Dominican Republic	19.7	–	21.7	25.0
El Salvador	19.4	22.3	–	26.6
Grenada	–	–	45.3	–
Guyana	21.5	–	24.4	–
Guatemala	19.7	21.3	–	16.9
Haiti	22.4	23.8	30.0	–
Honduras	20.0	–	–	20.4
Jamaica	30.0	–	38.0	–
Mexico	21.2	–	15.2	17.3
Nicaragua	20.2	–	–	24.3
Panama	20.4	21.9	21.5	22.3
St Kitts & Nevis	–	–	45.6	–
St Lucia	28.7	–	38.8	–
St Vincent	–	–	42.4	–
Trinidad and Tobago	22.1	–	25.3	–

Sources: González de la Rocha (1999: 153); UN (1995: 33–37); UNDP (1995: 63–5).
Notes
* Data for these years or nearest available year
– = no data

black women were targeted for sterilisation (Sarduy and Stubbs 1995). In Mexico, in turn, the greatest take-up of family planning was among middle-income groups with the poor tending to eschew contraception (Chant 1999a).

Women's rising age at marriage is another factor which has depressed fertility.[4] Although data on this are rather patchy (see Table 2.2), and wedlock is not necessarily a prelude to having children, later age at marriage, often thought to have been stimulated by women's rising education and labour force participation, may delay fertility and thus contribute to reducing birth rates. In Mexico, for example, the median age at marriage for the population as a whole increased from 20.8 to 22 years between

---

[4.] This is not universally the case in the Middle American region. Citing data from the Latin American Demographic Centre, Arriagada (1998: 86n) notes that there has actually been a lowering in the age of first union in Haiti and Jamaica. Recent research on St Vincent also indicates that while marriage was often the end stage rather than the beginning of a long-term co-resident union, marriage rates are presently increasing rather than falling (Chant and McIlwaine 1998: 12).

1970 and 1990, largely due to a decline in early marriages (López and Izazola 1995: 12). In turn, the average size of family-based households (male- and female-headed) fell from 5.4 to 5 between 1960 and 1990 (ibid.: 15).

Another influence identified as important in reducing household size is the growth in number of single-parent families, most of which are headed by women (UN/ECLAC 1994: 76; see also Box 2.2). Yet although this may conceivably have had some effect, it is important to recognise that these units do not necessarily consist of mothers living alone with their children, with female-headed households having a much greater likelihood of being extended than their male-headed counterparts (Arriagada 1998; Chant 1997). Moreover, it should be noted that establishing numbers of lone parent- and/or female-headed households is notoriously difficult. Data on female household headship are often inaccurate and underestimated, not to mention occluded by vague definitions which vary across countries (Chant 1997). When it comes to establishing trends in household headship over time, complexities are further compounded by the fact that census bureaux in the bulk of Middle American countries have only recently begun to register gender breakdowns of household headship. None the less, official data for the last decades of the twentieth century point to increases in most parts of the region. In Guyana, for example, women-headed households grew from 22.4 per cent to 35 per cent of the population between 1970 and 1987 (Patterson 1994: 122). Between 1980 and 1990 they rose from 17.6 per cent to 20 per cent in Costa Rica, from 21.5 per cent to 22.3 per cent in Panama, and 21.7 per cent to 25 per cent in the Dominican Republic (UNDP 1995: 63–5; see also Table 2.2). Currently, over 20 per cent of households in Central America are headed by women, and in the Caribbean 40 per cent or more (Arriagada 1998: 91).

These tendencies have fed into a general consensus that the 'patriarchal family model', thought to have prevailed in earlier historical periods, is on the wane. The growing incidence of lone motherhood and female house-hold headship has been accompanied by, and embedded in, falling levels of legal marriage, rising numbers of out-of-wedlock births, greater rates of divorce and separation, and mounting involvement of women in the historically male preserve of family 'breadwinning' (Arriagada 1998; Benería 1991; Cerrutti and Zenteno 1999; Folbre 1991; Jelin 1991; González de la Rocha, 1995). In Costa Rica, for example, official figures indicate that the proportion of births outside marriage had increased to 44 per cent by 1994, from a level of 37 per cent in 1980 (MIDEPLAN 1995: 5–6). Over the same period, marriage rates fell from 30.8 to 23.5 per 100 (ibid.), and between 1980 and 1996 the number of divorces rose from 9.9 to 21.2 per 100 (PEN 1998: 210). More recently, a report by the Instituto Nacional de la Mujer (National Institute for Women) (INAMU 1998) revealed not only that one in four Costa Rican children had a *padre desconocido* (unknown father), but that teenage motherhood was on the rise, with 60 per cent of single parents in the country being under 25 years of age, and 16 per cent under 18 years. Interestingly, many of these changes parallel those in other parts of Latin America (Kaztman 1992), and elsewhere in the world (see Castells 1997; Moore 1994).

Factors contributing to household diversity in contemporary Middle America

Recognising that Middle America's highly variegated patchwork of household types is due in part to historical legacies, important influences on contemporary household dynamics still to be discussed include migration, social and legal changes around marriage and divorce, and neoliberal economic restructuring.

Migration and households

Migration has played a highly significant role in stimulating changes in household structure in Middle America in the postwar period. Linked to increases in urban economic development and the decline of subsistence production in rural areas, households have fragmented and reconstituted themselves wholly or partially as a result of the movement of members to urban areas, whether on a short- or long-term basis.

One particularly critical aspect of migration with regard to household dynamics is the female selectivity of long-term or permanent rural–urban movement in Middle America. The dominance of women in rural–urban migration flows is partly a function of the juxtaposition of declining productive possibilities for women in the countryside, and relatively greater income-earning opportunities in towns. Expansion of female employment opportunities has occurred mainly in domestic service, and, in places such as the northern border of Mexico, Puerto Rico and the Dominican Republic, in multinational export-processing manufacturing (Fernández-Kelly 1983; Safa 1995a, 1999). Many studies have also shown that female-selective migration is both a cause and consequence of female household headship. Not only is the female bias in migration flows responsible for a 'feminisation' of sex ratios in towns, thereby making the establishment of female-headed households more likely, but female household heads are an important sub-group of women migrants from rural areas, especially where lack of land, labour supply or employment opportunities renders them unable to support their dependents (see Bradshaw 1995 on Honduras; Chant 1997 on Mexico and Costa Rica; Yudelman 1989 on Central America in general).

Female-selective rural–urban migration exerts an impact not only on the structure of households in towns, but also in rural areas, as shown by Gloria Rudolf's (1999) longitudinal study of the highland village of Loma Bonita in Coclé province, Panama. In the wake of an upsurge of rural–urban movement in the early 1960s, three times as many women as men left Loma Bonita, and even though this differential had declined somewhat by the period 1966–1972, it was still almost double (Rudolf 1999: 107). Although many female migrants were young, single and childless, lack of opportunities in the village meant that increasingly lone mothers too had to uproot and leave their children behind, thereby leading to the evolution of grandparent- and/or grandmother-headed households (ibid.: 119).

Grandmother-headed households have been documented in the Caribbean over a longer time period, as part of a vast and varied constellation of female-headed and female-centred domestic arrangements (Momsen 1992; Smith 1988). Alongside slavery and the perpetuation of West African kinship practices in the diaspora (Blumberg and García 1977; Williams 1986), international migration has traditionally been deemed a major factor responsible for female household headship in the area. In Barbados, for example, male selectivity in overseas migration was the major catalyst for female household headship during the first part of the twentieth century. It declined in importance only when increased numbers of Caribbean women joined the ranks of overseas migrants to Britain and Commonwealth countries such as Canada, in the 1950s, sometimes as part of family reunification programmes, but often as workers in their own right. The persistence and increase in levels of female headship since the mid-twentieth century have been attributed to high rates of female labour force participation, greater rights and self-assertiveness on the part of women, and lower rates of legal marriage (Chant and McIlwaine 1998: 56). As for mainland Middle America, however, gender selective rural–urban as well as international migration continues to be critical in the establishment of female-headed households (Acosta-Belén and Bose 1995: 25). Here men remain dominant in international flows, with the US being the most important destination, particularly for Mexican migrants. Although undocumented international migration precludes precise calculations (Diaz-Briquets 1989: 33), by 1990 an estimated 25 million people in the US were 'Hispanic', with around half being of Mexican origin (Green 1991: 59). Dominicans are another sizeable group in the US. Sørensen's (1985) study of Dominican migration, for example, reveals that about 15 to 20 per cent of the population has been involved in migration in recent years, with between 500,000 and one million living in the US (mainly in New York City).

As for Central America, steep rises in international migration in the late 1970s were driven by war and conflict in Nicaragua, El Salvador and Guatemala. Between 1980 and 1990, El Salvador accounted for 40 per cent of regional emigration with nearly half a million people (PER 1999: 363). Many Salvadorean refugees went to the US, as did those from Guatemala, although some of the latter (often women) also took up residence in refugee camps in southern Mexico. Nicaraguan refugees tended to move south into Costa Rica (PER 1999: 363); this resulted in an estimated 4.5 to 7.5 per cent of Costa Rica's population in the mid-1980s being from outside the country (Diaz-Briquets 1989: 38). Although some of the immigrants were political refugees, however, many also had migrated to Costa Rica for economic reasons such as greater availability of jobs or lower costs of living (Chant 1992). At the same time, labour migration from Costa Rica is also noted, especially among Afro-Caribbeans in Limón on the Atlantic coast. Limited occupational mobility among this group (especially men) in a racially segmented labour market means that overseas migration (usually to work on cruise liners in Miami), is often the only route to enhancing their economic position (McIlwaine 1997). These migratory patterns have often split households and led to high levels of *de facto* female household headship, a tendency exacerbated by the fact that war and violence in many countries have created large numbers of widows (Acosta-Belén and Bose 1995: 25).

Having noted that long-term migration can fragment households, it should be emphasised that migrants frequently retain close ties with kin in home areas. Interactions between source and destination localities may endure even across long distances, and can comprise substantial monetary transfers. In the Dominican case, for example, the aggregate value of migrant remittances in the mid-1980s amounted to nearly as much foreign exchange as that generated by the country's sugar industry (Sørensen 1985). These remittances, in turn, are often a vital component of household livelihoods and testify to the importance of situating the analysis of households within their broader social networks.

Social and legal changes relating to marriage and divorce

Another set of factors with important outcomes for household structure and organisation in Middle America lies in the sphere of legislative changes, particularly those which have strengthened women's position as individual citizens, or as family members, through interventions in family law, reproductive rights, sexual abuse and domestic violence, and labour market discrimination. Many of these changes have been driven by women's and feminist movements in the region (see for example, Craske 2000; Ewig 1999). In most countries in Middle America, for example, access to divorce on the part of women is now easier than in the past. This is not only a function of widened grounds for divorce, but also of measures to protect women's interests in the aftermath of conjugal breakdown; for example, in Costa Rica the amendment of Article 138 of the Family Code in the 1990s removed men's right to decide on the custody of children unless otherwise decreed by the courts (Vincenzi 1991). Yet although the number of divorces per 100 marriages in Costa Rica increased seven-fold between 1975 and 1991 (from 2 per 100 to 15.3 per 100), it is only one-third of the rate of Cuba, where half of all marriages now end in divorce. The disparity between Cuba and Mexico is even greater, with Mexico having only one-twelfth of the Cuban level (Lumsden 1996: 120). Thus although there is arguably increased social acceptance of divorce in Latin America in general (Durham 1991: 61), one must be wary of assuming that legislative shifts alone are sufficient. In the Mexican case, for example, LeVine (1993: 95) observes that: '. . . though nowadays there may be a good deal of talk about divorce, women who remain in seemingly untenable relationships still greatly outnumber those who get out'. In Costa Rica too, Fernández (1992) argues that divorce is still not acceptable to most people. Another factor is that divorce involves long drawn-out proceedings, with the costs being beyond the reach of many low-income women (Chant 1997: 137).

In order to avoid the expense, trauma and social stigma attached to the breakdown of a formalised, and particularly religious union, cohabitation is often seen as a more realistic alternative (see earlier). Although, traditionally, consensual unions have been less protected than formal marriages or have had no legal protection whatsoever (CRLP 1997: 12), the boundaries between marriage and cohabitation have diminished in a variety of Middle American countries in recent years. In Nicaragua, for example, the Sandinista regime introduced new laws that recognised stable, permanent unions as having the status of *acompañado*, or common-law marriage (Lancaster 1992: 18). In Costa Rica, the Social Equality Law passed in 1990 also made explicit recognition of free unions and gave rise to legislation that required compulsory registration of property in women's names in the event of separation (Molina 1993). Along with other initiatives that have followed in its wake, such as the National Plan for Equality of Opportunity Between Men and Women, the Programme for Female Heads of Household and the National Plan for the Prevention of Intra-Family Violence, this has expanded women's choices about whether or not to stay with partners, and conceivably made female household headship more tenable (Chant 1997; see also below).

Neoliberal economic restructuring

Last, but not least, the profound economic transformations that have occurred in Middle America in the 'Lost Decade' of the 1980s and beyond have also been associated with changes in household structure. As Radcliffe (1999: 197) notes for Latin America in general: '. . . the avocation of neoliberal development policies by most governments has significantly influenced the ways in which the nexus of labour–household–economy is organised, with consequences in turn for the nature of gender relations'.

Driven by World Bank and International Monetary Fund (IMF) Structural Adjustment Programmes (SAPs), stringency measures and the opening-up of Middle American economies in the 1980s and 1990s saw cutbacks in social expenditure, declining purchasing power of wages, rising levels of open unemployment, and informalisation of the labour market. The latter has comprised, on one hand, increased informal labour practices within the formal sector. In the face of pressure to reduce 'structural rigidities' in the workforce and to achieve greater 'flexibility', many firms have made greater recourse to subcontracting and piecework, as well as casualising the status of in-house employees so as to reduce obligations for social security contributions and fringe benefits (Benería and Roldán 1987; Green 1996; Miraftab 1994: 469; Peña Saint Martin 1996). Outside 'formal sector' firms, there has also been a rise in the number of people forced to create their own forms of livelihood in the absence of salaried alternatives. In Costa Rica during the 1980s, for example, the growth in numbers of informal sector workers was consistently higher than for wage-earners and employers (Valverde 1998: 53). In Guatemala City, as many as 80 per cent of new jobs in the early 1990s were informal (Green 1995: 107). In Cuba too, despite the traditional resistance of Castro's government to informal economic activity, the collapse of communism in 1990 gave rise to a massive expansion in this sector, now recognised by the state as making a contribution to economic development (Molyneaux 1996: 33). Although much of the growth in informal employment in Middle America is due to lay-offs both in private and public branches of the formal sector, not all refugees from the formal sector have been absorbed by the informal sector, as evidenced by a rise in open unemployment in Nicaragua from 4.5 per cent in 1986 to 23.5 per cent in 1994 (Bulmer-Thomas 1996: 326).

Even if women are still more likely to be unemployed than men (Radcliffe 1999: 201), it appears that men's work has been relatively harder hit by recession and structural adjustment in Middle America. This is mainly due to men's concentration in sectors such as heavy industry, construction and agro-export production which have been considerably more vulnerable to downswings in the macro-economic climate than more 'feminised' tertiary occupations such as domestic service. Another important factor is that women have been a favoured workforce in the labour-intensive *maquiladora* sector, which has been at the forefront of neoliberal strategies of export promotion (de Barbieri and de Oliveira 1989: 23; Ward and Pyle 1995). In some places, these developments have overlain long-term declines in male employment. In Puerto Rico, for example, the shift to urban-based industry and services and the disintegration of the sugar economy in the second half of the twentieth century led to significant employment losses for men (Safa 1995b). Between 1960 and 1981 labour force participation rates for men dropped from 75.9 per cent to 68.2 per cent, and although rates rose slightly during the 1980s (to 2.8 per cent), this was still lower than for women (4.2 per cent) (Safa 1999: 2).

Notwithstanding positive trends in female economic activity prior to the outbreak of crisis in Middle America, in part due to increases in education, and in part due to the expansion of services, commerce and export-processing manufacturing, for low-income women in particular, economic hardship seems to have been the decisive factor in stimulating the upsurge of the last two decades (see also Benería 1991; Chant 1996; González de la Rocha 1988). As Safa (1995b: 33) summarises, the crisis has increased the 'importance and visibility of women's contribution to the household economy as additional women enter the labour force to meet the rising cost of living and the decreased wage-earning capacity of men'. This is reflected not only in the general rise in women's employment, but also in increased numbers of older and/or married women workers. In Mexico, for example, over one-quarter of married women were recorded as working in 1991 compared with only 10 per cent in 1970, and the highest levels of economic activity are now in the 35- to 39-year age cohort (43 per cent) (CEPAL 1994: 15; González de la Rocha 1991: 117). In Costa Rica too, while there was only one female worker for every three men in the 20- to 39-year age cohort in 1980, by 1990 the gap had narrowed to one in two (Dierckxsens 1992: 22).

To some extent, women's entry into the labour force has compensated for men's deteriorating incomes, yet because women's earnings are so much lower than those of men, this has not by any means protected households from inflationary increases in costs of living. This is exacerbated by the fact that women are clustered in informal economic activity and part-time work[5] (see McIlwaine *et al.*, this volume for discussion of the reasons

[5.] In 1994, for example, women's average wage rates for non-agricultural employment in Costa Rica were only 83 per cent of those of men, with corresponding figures for Mexico being 75 per cent (UNDP 1995: 36). For Latin America and the Caribbean as a whole, women on average earn only 67 per cent of male wages (CRLP 1997: 12). The significance of the informal sector in these disparities is that in the case of Central America, the gender earnings differential in informal employment averages 25 per cent, whereas in the formal sector this is only 10 per cent (Funkhouser 1996: 1746).

for this). In addition, the rising precariousness of employment and loss of fringe benefits which subsidised healthcare, or which provided cover for illness and accident, and so on have transferred extra costs on to low-income families. A now seemingly long-term situation of financial scarcity and employment insecurity is observed to have weakened solidarity among household members as a result of stress, frustration and despair, even when they are not forced physically apart through migration (see González de la Rocha 1997). On top of this, although women's income-generating work may have cushioned the impact of crisis and restructuring in economic terms, ideologically and psychologically, men's loss of employment and their mounting dependence on women's earnings has threatened masculine identities (Gutmann 1996). In light of this, it is perhaps no surprise that there has been an intensification of domestic strife during the crisis period (Benería 1991; Gledhill 1995: 137; González de la Rocha *et al.*, 1990; Safa 1999). As Selby *et al.* (1990: 176) argue:

> Male dignity has been so assaulted by unemployment and the necessity of relying on women for the subsistence that men formerly provided, that men have taken it out on their wives and domestic violence has increased . . . the families which have been riven by fighting and brutality can easily be said to be the true victims of economic crisis. . . .

In some cases conflict has been enough to break up households, and not necessarily through men walking out. Indeed, women's increased economic independence has often been seen as having strengthened women's ability to decide to terminate relationships, with the rise in women's labour force participation often being linked with the 'feminisation' of household headship in Middle America (see for example, Bradshaw 1995; Chant 1997; Safa 1995a, 1999).[6] It is clear, therefore, that gender occupies a crucial place in the analysis of changes in households in the late twentieth century, as discussed in greater detail below.

Gender identities and households in Middle America

As little as 20 years ago, most writings on gender in Central America, Mexico and the Hispanic Caribbean emphasised polarity between men's and women's roles within the household. The picture painted of 'traditional' familial gender patterns was one which revolved around men as primary (if not exclusive) breadwinners, the main (if not sole) decision-makers within household units, and as possessors of considerably greater power and freedoms than their female counterparts. Women, on the other

---

[6.] While it is undoubtedly the case that access to employment enhances women's scope to head their own households, there is not necessarily a direct or consistent relationship. For example, in their review of Mexican urban areas, Cerrutti and Zenteno (1999: 72) point out that despite a significant rise in female labour force participation between 1988 and 1997, there was only a negligible increase in the proportion of households headed by women over this period.

hand, tended to be portrayed as mothers and housewives, dependent on men financially, and with few powers of determination over their own lives. Basic divisions between men and women with regard to labour, power and resources within the home were also (and in many cases still are) seen as accompanied by, and embedded in, polarisations in morality, sexuality and social behaviour in which the limits of female activity were confined to the domestic domain. Although the conceptual underpinnings of such dualistic stereotypes have since been challenged by recognition of gender's variegation by class, 'race', age and locality, and by concerns to rescue the lives of individual men and women in the South from blanket ethnocentric constructions, it is also the case that economic, social and demographic changes in the last two decades have played an important part in reducing gender divisions.

The impacts of female labour force participation on gender identities and households

As mentioned in the previous section, one of the major effects of economic restructuring has been the rise in female labour force participation. This seems intimately bound up with the increased prevalence of female-headed households, both of which, in turn, are viewed as having had positive outcomes for women's autonomy and 'empowerment' (Chant 1997; McClenaghan 1997; Safa 1995a). In my own studies of low-income households in the Mexican cities of León, Querétaro and Puerto Vallarta, for example, there seems to be greater collective negotiation and scrutiny over financial affairs where households have diversified their income strategies, and becoming a female head is less daunting where women are already earning (Chant 1991). As echoed by Cerrutti and Zenteno (1999: 71) in the wider Mexican context, rising female labour force participation 'has had profound repercussions in the breaking of the traditional model of male head as the sole economic breadwinner in the household' (author's translation). With reference to Puerto Rico, Cuba and the Dominican Republic, Safa (1995a: 58) also notes that although the cultural norm of the male breadwinner remains decidedly embedded in the workplace and the state, 'women's consciousness and increased bargaining power in the household have increased as a result of the massive incorporation of women into the labour force in recent decades'. Safa (1995b: 33) further observes that women's declining dependence on male incomes and growing economic participation in their own right have presented a major challenge to the 'myth of the male breadwinner' at the grassroots.

Despite the transformative potential of these changes, however, it is important to bear in mind that women are still far from 'free agents' when it comes to taking employment. They are hampered by lower levels of education and vocational training than men, and their availability for work is tempered by their domestic and childrearing responsibilities (González de la Rocha 1994: 141–2). Not only may their choice of employment be subject to approval from husbands, but as Townsend *et al.* (1999: 38) note for Mexico, they may need their husbands' permission to work at all. Sometimes such prescriptions are enshrined in civil law. For example, the

Guatemalan Civil Code states that a woman has a 'special right and duty to nurture and care for her children' and can only be employed outside the home if this does not go against the children's interests or undermine the needs of the household (Steiner and Alston 1996: 890). Moreover, husbands have the right to prohibit the work of wives if they themselves provide sufficient income, however that might be determined (CRLP 1997: 20). González de la Rocha (1994: 141–2) further argues with reference to her research on Guadalajara, Mexico, that women's employment has failed to provoke more in the way of change in domestic power relations because their earnings are so low, and because they cannot necessarily control their own earnings.

Questions of gender identity also enter into assessments of the extent of change in gender regimes. McClenaghan (1997: 29) notes for the Dominican Republic, for example, that even where women are the primary providers, men are usually still acknowledged as *el jefe* (head of household). This is partly because men's ownership of land and property places them in a position of authority regardless of whether they are actually employed (see also Safa 1999: 6 on Alta Gracia in the Dominican Republic). Moreover, as far as women themselves are concerned, it is by no means clear whether their rising involvement in income-generating activity has translated into major shifts in identity or empowerment. On one hand, Cerrutti and Zenteno (1999: 71) claim in the Mexican context that:

> ... the increase and diversification of women's labour experience has implied an erosion of prescriptive norms and their roles, particularly in relation to the ideology of reproduction [author's translation].

Yet García's and de Oliveira's (1997) study of low-income and middle-class women in the Mexican cities of Tijuana, Mérida and Mexico City finds that for low-income women in particular, motherhood remains their primary source of identity, with employment viewed merely as an instrumental means by which they can better fulfil their mothering roles (García and de Oliveira 1997: 368). Only among middle-class mothers, who are able to afford paid childcare and who have sufficient education to enter non-manual occupations, is work more likely to be a significant aspect of personal identity (ibid.: 381). One plausible reason is that educated middle-class women are more likely to have long-term employment rather than intermittent involvement in the workforce (Cerrutti 2000). Another is that most middle-class women are able to pass on basic reproductive tasks to domestic servants. For low-income women, however, housework and childcare remain firmly in their domain. This means a heavy double burden of labour when they take up employment that arguably leaves little time for reflection on how their critical efforts in household survival might be a route to more egalitarian gender relations (Chant 1996: 298).

Men and reproductive labour

Leading on from the above, lack of fundamental change in gender is reflected in, and perpetuated by the fact that men have rarely taken on greater shares of housework and childcare to compensate for their spouses' increased participation in income generation (Chant 1996: 298). Although Gutmann (1999: 167) claims that in Mexico young husbands and fathers are doing more childcare than their elders, in Panama Rudolf (1999) argues that there was more gender complementarity in reproductive as well as productive labour in the past, at least in rural areas. This was possibly related to greater overlap between reproductive and productive spheres in subsistence farming areas. For the Dominican Republic, Safa (1999: 16) reports that even where men are unemployed, they tend not to switch the time freed up to housework and childcare. This is also the case in Cuba, regardless of the fact that the Family Code of 1975 prescribed that men should assume an equal share of work in the home. As Pearson (1997: 677) asserts with reference to Cuba: 'In spite of official desires to dissolve the social division of work by gender, the redistribution of reproductive work between men and women was in fact limited.' Pearson goes on to observe that during the economic crisis of the 1990s: '. . . the pressures to maintain household consumption levels in increasingly difficult circumstances tended to reinforce the traditional gender division of labour rather than resolve it' (ibid.: 700). This is mainly because of the additional work involved in reproductive labour arising from cutbacks in state services, and the time and care that needs to be taken to secure the fulfilment of household members' basic needs. In short, women are often reluctant to let their spouses waste precious resources through lack of skill or experience. Women may also be reluctant to let their men participate in housework or childcare because it suggests that they do not have a 'real man' for a partner. Indeed, traditional gender stereotypes are often perpetuated by both sexes among their peers, as well as through the socialisation of children (see Lancaster 1992: 44).

Men in crisis?

Part of men's apparent unwillingness to participate in reproductive labour conceivably stems from desires to protect the remaining vestiges of a 'masculine identity' in a world in which women's activities are widening, not to mention encroaching upon 'male territory'. Refusal to assist women could therefore be interpreted as passive resistance to a situation which threatens the gender status quo. As Townsend *et al.* (1999: 29) observe for Mexico:

> In the 1980s, many men lost the possibility of being a good provider, and lost much of their power and status outside the home. The patriarchal bargain came under threat. Women could still go on suffering (not a virtue in a man): men could no longer be real men, but women were still real women. In both urban and rural areas, many men took out their frustrations and loss of power outside the home within it, with appalling consequences for women and children.

Aside from non-participation in reproductive labour, resistance to change has been noted in other forms; for example, a common male 'backlash' to women's enhanced income-earning opportunities is for men to withdraw their own financial support from the household (Chant 1997; Safilios-Rothschild 1990). Similarly, inability to provide can drive men away from 'paternal responsibilities' (Engle 1997: 35). At a more extreme level, men's frustrations can also take the form of outright violence.

Although men in most Middle American countries have retained a larger share of employment and earnings, their mounting fragility as primary breadwinners seems to be the central issue in the increasingly widely noted 'crisis of masculinity' in the region (Chant 2000; Escobar Latapí 1999; Gomáriz 1997). Within the context of conjugal households, this has tended to undermine men's security about their position and privileges, about dependency and allegiance on the part of wives and children, and ultimately about their own power to determine the course of family unity and/or disunity (Chant 2000). As Safa (1999: 8) observes for the Dominican Republic: 'Marital life still consists of a succession of consensual unions, but now the initiative for break up rests as much with women as with men.' Indeed, while men have often been rather peripatetic figures in Middle American households, this was not really a source of concern so long as men themselves could rely on the family being there when they needed it. Now that more women are earning and have greater bargaining power and independence, these guarantees have been toppled. Indeed, in the author's study on 'family crisis' among male and female youth and adults in Costa Rica (see Note 1), it was men who voiced most concern about the demise of the 'traditional' family, and a large share of the blame was placed on women's increased absence from the home through employment (Chant 1999b). Similar observations have been made in other parts of Middle America such as Barbados (Chant and McIlwaine 1998: Chapter 3). With women being held responsible for 'family breakdown', an obvious implication is that women rather than men should do something to change the course of events.

Trends towards men's increasingly precarious position within conjugal households are widely perceived to have been exacerbated by legislative and policy initiatives in women's interests. Although in principle, most men in Guanacaste, Costa Rica, claim to be in favour of gender equality, and express distaste for what they feel are outmoded *machista* attitudes, considerable concern is voiced over the potential 'abuse' by women of new entitlements accorded through recent government programmes and laws (see earlier). As one 25-year-old construction worker whom the author interviewed in 1997 put it, the Law for Social Equality *'no funciona'* ('doesn't work') because it encourages women to think they are superior to men, and many women *'quisieran ser más del hombre'* ('want to be more than men'). This was echoed by a respondent in his sixties who said that although he thought it was good that there was more equality between women and men nowadays, that *'a veces, sobrepasa la cosa y la mujer regaña al hombre!'* ('at times this goes too far, and women tell men off!'). He felt that new laws had made it difficult for men to *'volar haches'* ('wield the axe'), meaning 'to exert control', but with overtones of domestic violence, which has recently become more likely to meet with imprisonment than in the past (Chant 2000). Indeed, while *machismo* is often argued to be an outdated stereotype, there are many instances in which this powerful complex of male power and superiority, so frequently associated with violence against women (see Chávez 1999; Montoya Tellería 1998), is rearing its head as signs of a new, more egalitarian, gender regime struggle tentatively into the twenty-first century.

State and civil society responses to household change: perspectives and policies

Having documented some of the major changes in gender and household organisation at the grassroots in Middle America, it remains to examine how state and civil society organisations have reacted to shifting household realities. Clearly, these reactions have been variable across groups as well as over time. In the space of a single decade, for example, the Roman Catholic Church voiced resistance to a programme for lone mothers in Honduras (Grosh 1994: 84–5), but was actively involved in selecting candidates for a similar programme in Costa Rica a few years later (Chant 1999a). By the same token, it is possible to see, in general terms, that resistance to accepting family transitions is diminishing in many parts of the region. In the early 1980s, for example, the Nicaraguan women's organisation, AMNLAE, was charged by its political opponents with 'mounting a communistic attack on the sanctity of the family', in the wake of its calls for legislation to reformulate the family as an 'institution not of *machismo* and patriarchy but of equality and responsibility', and to promote recognition and acceptance of household diversity (including high numbers of female-headed households) (Lancaster 1992: 17–18). In the 1990s, however, a different mood prevailed, with the case of ex-President Daniel Ortega's alleged sexual abuse of his stepdaughter having brought condemnation from men and women in all sectors of Nicaraguan society, and with the establishment of ground-breaking efforts to address the issue of gender-based intra-family violence (see also below).

While in general terms there is clearly ongoing anxiety about the 'breakdown' of the patriarchal nuclear family, therefore it seems equally apparent that some change within conjugal households is deemed necessary to prevent further disintegration and, for those households, predominantly low-income ones, who have already 'broken out' of the conventional mould, to provide assistance that better ensures their security and stability (see Guéndel and González 1998: 28). These imperatives are especially pertinent in an era of state cutbacks and the aggressive marketisation of national economies in which family cohesion is likely to have to be relied upon as a major bulwark against growing poverty.

In Costa Rica, for example, interviews conducted with a range of government officials and NGO representatives in 1999 indicated some willingness to disregard traditional moral objections to declining rates of marriage and rises in divorce and separation, and to take pragmatic steps towards adjusting to greater diversity in family patterns over time. This seems to result from the fusion of local consciousness of family realities, with the presence in San José of the many regional headquarters of international organisations, such as UNICEF, which have launched a variety of progressive social initiatives (ibid.; see also below). This is reflected in the increased targeting of social assistance to 'vulnerable groups' such as lone parents and adolescent mothers, and in attempts to strengthen the protection and rights of children. For example, the 'Solidarity Plan' of Costa Rica's current Social Democratic government (1998–2002), which aims primarily to reduce poverty, has geared three of its six main strategic areas to family support, one being to strengthen family cohesion, one to assist women in extreme poverty (with priority accorded to female heads of household), and one to providing assistance for children and youth. This initiative is accompanied by new legislation for low-income women and youth, and the extension of the activities and responsibilities of specialist

Plate 2.1 Women's Association (*Asociación de Mujeres*), Guatemala City: overcoming the odds together (*superandonos juntas*)
Photo: Cathy McIlwaine

organisations such as the National Child Welfare Institute. Beyond this, the work of the Catholic Church and international agencies such as UNICEF and the United Nations Institute for the Prevention of Crime and the Treatment of Delinquency (ILANUD) are complementing the programmes of the Costa Rican state with interventions to strengthen family unity, to promote the human rights of women and children, and to effect reductions in child abuse and domestic violence. These initiatives are not confined to Costa Rica, as illustrated in relation to two specific examples of these recent types of intervention below.

Public intervention to reduce domestic violence

Although domestic violence has existed in most societies throughout history, from the 1980s onwards the rise of groups demanding women's rights in Central America has shone a spotlight on the problem (Claramunt 1997: 96). Although data remain scarce and unreliable, a report by the Panamerican Health Organisation from the early 1990s suggested that around 50 per cent of Guatemalan women suffered from domestic violence (ibid.). In Nicaragua, where there have been the most systematic efforts made to record and respond to domestic violence, a study conducted in León, Nicaragua's second largest city, in 1996, with 488 women of 15 years or more, revealed that 52 per cent of women had been the victim of some kind of conjugal violence in their lives, and of these 27 per cent had experienced physical assault by their partners in the year leading up to the study, which in 70 per cent of cases were catalogued as severe (ibid.).

Reflecting the seriousness of the problem, most countries in Middle America now have in place some form of legislation and intervention against domestic violence. In Barbados, for example, a law on protection orders against domestic violence was passed in 1992, and victims of domestic violence now have access to police support and telephone hotlines. There is also special training for police officers to deal with domestic violence (Arriagada 1998: 94). Jamaica's law on domestic violence was passed in 1996, and state actions have included preventive campaigns in schools and the provision of shelters for victims of domestic abuse. In Honduras, the law for the prevention, punishment and elimination of violence against women was passed in 1997, and in addition to legal aid for women, workshops have been organised to increase awareness of the problem (ibid.). Nicaragua, whose law against intra-family violence was passed in 1996, stands out as one country in the region where men have organised to eliminate domestic violence, and in order to do so, are attempting to 'unlearn' their *machista* socialisation (Montoya Tellería 1998). Aside from men's own efforts to put a brake on violence, Gutmann (1997) argues that men who have reacted violently towards women can and do change as a result of women's instigation as well.

The place of female-headed households in social policy

As mentioned earlier, another notable element in social policy in recent years has been the increased attention accorded to female-headed households.

Although this has not been quite as widespread in the region as initiatives to reduce domestic violence, it is quite likely that a general shift from universal to targeted social programmes will lead to more programmes for female-headed households in the future, especially given growing emphasis on child rights and protection.

In Costa Rica, the Social Welfare Ministry established its first programme for female household heads in 1997 during the regime of President José María Figueres (1994–98) of the more left-wing of Costa Rica's two main political parties. This offered a temporary stipend for women to enable them to take courses in assertiveness and skills training. While arguably limited in its coverage and impacts (see Budowksi and Guzmán 1998), the programme has been extended into the new administration of Miguel Angel Rodríguez (1998–2002) under the auspices of the *Programa de Atención para las Mujeres en Condiciones de Pobreza* (Programme for Women in Con- ditions of Poverty). This has been accompanied by two further programmes aimed at the young. The first of these *Amor Joven* (Young Love), launched in 1999 by the First Lady, Lorena Clare de Rodríguez, is concerned with preventing adolescent pregnancy; the second is entitled *Construyendo Oportunidades* (Building Opportunities), and seeks to (re)integrate young mothers into education and to provide special programmes of state support for their children (see Chant 1999a, 2000).

Other countries in Middle America that have launched programmes specifically targeted at women-headed households include Honduras and Puerto Rico (see Grosh 1994; Safa 1995a). Even if elsewhere in the region governments are more hesitant about promoting schemes that may increase numbers of female-headed households, other forms of support for women can have similar outcomes. In Cuba, for example, there is no pro- gramme that provides special welfare benefits to female heads, but poli- cies favouring greater gender equality, higher levels of female labour force participation, and the provision of support services such as daycare have all made it easier for women to raise children on their own (Safa 1995a).

Working with men

While attempts to strengthen women's abilities to support their children independently in Middle America have been welcomed in a number of circles, the challenge also exists to find ways in which they might be assisted by men and how men might be better integrated into family life, especially in the light of their apparently increasing marginality and the current dilemmas over changing gender roles and identities. One major priority identified by feminist writers is that of promoting greater co- operation in household responsibilities (see Arriagada 1998: 97).

In many respects, the legislative bedrock for shared parenting and family responsibilities is already in place, with a range of Family Codes in the region, for example in Cuba, El Salvador and Mexico, dictating that both men and women should take responsibility for financial provi- sion, housework and childcare (CRLP 1997: 102 and 160). The fact that the 'male breadwinner/female mother-housewife' model continues to be such a powerful normative ideology, and that in practice it allots women even more work than in the past, however, suggests that measures for more effective monitoring of legal prescriptions would be desirable. On top of this, although women continue to be identified as the primary carers of children, and there is generally little discussion of new roles for men within family life, some imaginative proactive approaches to redressing gender imbalances are beginning to emerge. Although in their infancy, plans are being mooted to include men in programmes of gender 're-socialisation' in Costa Rica (see Chant 2000), and Engle (1997: 35) discusses the establishment of fathers' groups by the Caribbean Child Development Centre based in Kingston, Jamaica, which have formed an organisation called 'Fathers Inc.'. In these groups, fathers who often do not share a home with their children learn parenting skills. Reasons given for the success of the groups are that they are men-only, and are initiated by men's interest in their children (ibid.). Indeed, evidence suggests that these social policy initiatives geared to 'bringing men back in' to family life may not be as difficult as anticipated given a growing body of research revealing men's existing deeply held concerns about their primary kin relationships. In Guatemala, for example, focus groups have shown that men express a profound sense of paternal responsibility for their children, even if in practice some are not in contact with them (see Engle and Alatorre Rico 1994: 4).

Concluding comments: gender and family in the twenty-first century

While a number of trajectories in gender and 'the family' in Middle America seem possible, the most likely is that kin relationships will continue to form the core of social organisation, but in a greater variety of forms, and with more flexible positioning of the members of household units with regard to rights and responsibilities. Yet whether households that depart from traditional normative ideals of structure and internal divisions of labour will receive the same ideological and institutional legitimacy as others is arguably less certain. Even in Costa Rica, where there have been important shifts in accepting 'new' household types, common reference in public and policy circles to factors such as 'family disintegration', and to the ways in which lone parenthood can disadvantage the material and emotional welfare of children and prevent their social integration would suggest that there remains, covertly if not overtly, an idealisation of the 'traditional' two-parent patriarchal family unit. In many ways this is ironic, given a groundswell of evidence not just in Costa Rica, but elsewhere in the region, that women-headed households are not necessarily the 'poorest of the poor' (Chant 1999a on Costa Rica and Mexico; Gafar 1998: 605 on Guyana), nor are their offspring necessarily likely to suffer in respect of nutrition, healthcare, education and so on (Engle 1995 on Guatemala). Insufficient recognition is also made of the fact that support networks of kin often greatly aid survival in women-headed households (González de la Rocha 1994; Safa 1999; Willis 1993).

If policymakers and other gatekeepers are able to draw from and help consolidate the rich range of survival mechanisms already adopted by women, men and their households at the grassroots, then there may be hope for greater gender equality among future generations. Acknowledgement, acceptance, and the promotion of greater public tolerance of household diversity are likely to be one of the most critical elements in this struggle.

Further reading

Bose, Christine and Acosta-Belén, Edna (eds) (1995) *Women in the Latin American Development Process*. Temple University Press, Philadelphia.
A collection of papers which reviews different dimensions of gender and development in Latin America with reference to various Middle American countries such as Mexico, Guatemala and the Dominican Republic.

Chant, Sylvia (1997) *Women-headed Households: Diversity and Dynamics in the Developing World*. Macmillan, Basingstoke.
Includes detailed case studies of women-headed households in Mexico and Costa Rica based on in-depth primary field research, within a more general review of the evolution and characteristics of these households in developing regions and at a global level.

Gutmann, Matthew (1996) *The Meanings of Macho: Being a Man in Mexico City*. University of California Press, Berkeley.
In a study based on first-hand fieldwork in a low-income neighbourhood of Mexico City, Gutmann explores the much-used, overly stereotyped and often poorly understood concept of *machismo* and evaluates its relevance to the lives and behaviour of local men.

Jelin, Elizabeth (ed.) (1991) *Family, Household and Gender Relations in Latin America*. Kegan Paul International/UNESCO, London/Paris.
A collection of papers on different aspects of household and family in Latin America, many by Latin American authors, including analyses of different historical periods and of different socio-economic groups.

Safa, Helen (1995) *The Myth of the Male Breadwinner: Women and Industrialisation in the Caribbean*. Westview, Boulder, Col.
A detailed comparative study of women's work and its implications for household change in Cuba, Puerto Rico and the Dominican Republic in the second part of the twentieth century, based on Safa's extensive fieldwork in the Hispanic Caribbean.

References

Acosta-Belén, E. and **Bose, C.** (1995) Colonialism, structural subordination and empowerment: women in the development process in Latin America and the Caribbean, in Bose, C. and Acosta-Belén, E. (eds) *Women in the Latin American Development Process*. Temple University Press, Philadelphia: 15–36.

Arriagada, I. (1998) Latin American families: convergences and divergences in models and policies, *CEPAL Review* 65: 85–102.

de Barbieri, T. and **de Oliveira, B.** (1989) Reproducción de la fuerza de trabajo en América Latina: algunas hipótesis, in Schteingart, M. (ed.) *Las Ciudades Mexicanas en la Crisis.* Editorial Trillas, Mexico DF: 19–29.

Benería, L. (1991) Structural adjustment, the labour market and the household: the case of Mexico, in Standing, G. and Tokman, V. (eds) *Towards Social Adjustment: Labour Market Issues in Structural Adjustment.* International Labour Organisation, Geneva: 161–83.

Benería, L. and **Roldán, M.** (1987) *The Crossroads of Class and Gender: Industrial Homework, Subcontracting and Household Dynamics in Mexico City.* University of Chicago Press, Chicago.

Blumberg, R.L. with **García, M.P.** (1977) The political economy of the mother–child family: a cross-societal view, in Leñero-Otero, L. (ed.) *Beyond the Nuclear Family Model.* Sage, London: 99–163.

Bradshaw, S. (1995) Women's access to employment and the formation of women-headed households in rural and urban Honduras, *Bulletin of Latin American Research* **14**(2): 143–58.

Brydon, L. and **Chant, S.** (1989) *Women in the Third World: Gender Issues in Rural and Urban Areas.* Edward Elgar, Aldershot.

Budowski, M. and **Guzmán, L.** (1998) Strategic gender interests in social policy: empowerment training for female heads of household in Costa Rica. Paper prepared for the International Sociological Association XIV World Congress of Sociology, Montreal, 26 July–1 August.

Buffington, R. (1997) Los jotos: contested visions of homosexuality in modern Mexico, in Balderston, D. and Guy, D. (eds) *Sex and Sexuality in Latin America.* New York University Press, New York: 118–32.

Bulmer-Thomas, V. (1996) Conclusions, in Bulmer-Thomas, V. (ed.) *The New Economic Model in Latin America and its Impact on Income Distribution and Poverty.* Macmillan in association with the Institute of Latin American Studies, University of London, Basingstoke: 296–327.

Castells, M. (1997) *The Power of Identity.* Blackwell, Oxford.

Centro Legal para Derechos Reproductivos y Políticas Públicas (CRLP) (1997) *Mujeres del Mundo: Leyes y Políticas que Afectan sus Vidas Reproductivas: América Latina.* CRLP, New York.

Cerrutti, M. (2000) Intermittent employment among married women: a comparative study of Buenos Aires and Mexico City, *Journal of Comparative Family Studies* **XXXI**(1): 19–43.

Cerrutti, M. and **Zenteno, R.** (1999) Cambios en el papel económico de las mujeres entre las parejas mexicanas, *Estudios Demográficos y Urbanos* **15**(1): 65–95. El Colegio de México, Mexico City.

Chant, S. (1991) *Women and Survival in Mexican Cities: Perspectives on Gender, Labour Markets and Low-income Households.* University of Manchester Press, Manchester.

—— (1992) Migration at the margins: gender, poverty and population movement on the Costa Rican periphery, in Chant, S. (ed.) *Gender and Migration in Developing Countries*. Belhaven, London: 49–72.

—— (1996) Women's roles in recession and economic restructuring in Mexico and the Philippines, *Geoforum* **27**(3): 297–327.

—— (1997) *Women-headed Households: Diversity and Dynamics in the Developing World*. Macmillan, Basingstoke.

—— (1999a) Women-headed households: global orthodoxies and grassroots realities, in Afshar, H. and Barrientos, S. (eds) *Women, Globalisation and Fragmentation in the Developing World*. Macmillan, Basingstoke: 91–130.

—— (1999b) Youth, gender and 'family crisis' in Costa Rica. Report to the Nuffield Foundation, London (October).

—— (2000) Men in crisis? Reflections on masculinities, work and family in Northwest Costa Rica, *European Journal of Development Research* **12**(2).

Chant, S. and **McIlwaine, C.** (1998) *Three Generations, Two Genders, One World: Women and Men in a Changing Century*. Zed, London.

Chávez, E. (1999) Domestic violence and HIV/AIDS in Mexico, in Foreman, M. (ed.) *AIDS and Men: Taking Risks or Taking Responsibility?* Panos Institute/Zed, London: 51–63.

Claramunt, M.C. (1997) *Casitas quebradas: el problema de la violencia doméstica en Costa Rica*. Editorial Universidad a Distancia, San José.

Comisión Económica para América Latina (CEPAL) (1994) El sector informal urbano desde la perspectiva de género: el caso de México. Paper presented at workshop 'El sector informal urbano desde la perspectiva de género', Mexico DF, 28–29 November.

Côrrea, S. with **Reichmann, R.** (1994) *Population and Reproductive Rights: Feminist Perspectives in the South*. Zed, London.

Craske, Nikki (2000) *Continuing the challenge: the Contemporary Latin American Women's Movement(s)*, Research Paper 23. Institute of Latin American Studies, University of Liverpool.

Diaz-Briquets, S. (1989) The Central American demographic situation: trends and implications, in Bean, F. Schmandt, J. and Weintraub, S. (eds) *Mexican and Central American Population and US Immigration Policy*. University of Texas Press, Austin: 33–64.

Dierckxsens, W. (1992) Impacto del ajuste estructural sobre la mujer trabajadora en Costa Rica, in Acuña-Ortega, M. (ed.) *Cuadernos de Política Económica*. Universidad Nacional de Costa Rica, Heredia: 2–59.

Dore, E. (1997) The Holy Family: Imagined Households in Latin American History, in Dore, D. (ed.) *Gender Politics in Latin America: Debates in Theory and Practice*. Monthly Review Press, New York: 101–17.

Durham, E. (1991) Family and human reproduction, in Jelin, E. (ed.) *Family, Household and Gender Relations in Latin America.* Kegan Paul International/ UNESCO, London/Paris: 40–63.

Ellis, P. (1986) Introduction: an overview of women in Caribbean Society, in Ellis, P. (ed.) *Women of the Caribbean.* Zed, London: 1–24.

Engle, P.L. (1995) Father's money, mother's money, and parental commitment: Guatemala and Nicaragua, in Blumberg, R.L., Rakowski, C.A., Tinker, I. and Monteón, M. (eds) *Engendering Wealth and Well-Being: Empowerment for Global Change.* Westview, Boulder, CO.: 155–79.

—— (1997) The role of men in families: achieving gender equity and supporting children, in Sweetman, C. (ed.) *Men and Masculinity.* Oxfam, Oxford: 31–40.

Engle, P.L. and **Alatorre Rico, J.** (1994) *Taller Sobre Paternidad Responsible.* Population Council/International Center for Research on Women, New York/Washington, D.C.

Escobar Latapí, A. (1999) Los hombres y sus historias: Society for Latin American Studies, Selwyn College, Cambridge, 9–11 April.

Ewig, C. (1999) The strengths and limits of the NGO women's movement model: shaping Nicaragua's democratic institutions, *Latin American Research Review* **34**(3): 75–102.

Fernández, O. (1992) Qué valores valen hoy en Costa Rica?, in Villasuso, J.M. (ed.) *El Nuevo Rostro de Costa Rica.* Centro de Estudios Democráticos de América Latina, Heredia.

Fernández-Kelly, M.P. (1983) Mexican border industrialisation, female labour force participation and migration, in Nash, J. and Fernández-Kelly, M.P. (eds) *Women, Men and the International Division of Labour.* State University of New York Press, Albany, NY: 205–23.

Folbre, N. (1991) Women on their own: global patterns of female headship, in Gallin, R.S. and Ferguson, A. (eds) *The Women and International Development Annual Vol. 2.* Westview, Boulder, CO: 69–126.

Funkhouser, E. (1996) The urban informal sector in Central America: household survey evidence, *World Development* **24**(11): 1737–51.

Gafar, J. (1998) Growth, inequality and poverty in selected Caribbean and Latin American countries, with emphasis on Guyana, *Journal of Latin American Studies* **30**(3): 591–617.

García, B. and **de Oliveira, O.** (1997) Motherhood and extradomestic work in urban Mexico, *Bulletin of Latin American Research* **16**(3): 367–84.

Gledhill, J. (1995) *Neoliberalism, Transnationalisation and Rural Poverty: a Case Study of Michoacán.* Westview, Boulder, CO.

Gomáriz, Enrique (1997) *Introducción a los Estudios sobre la Masculinidad.* Centro Nacional para el Desarrollo de la Mujer y Familia, San José.

González de la Rocha, M. (1988) Economic crisis, domestic reorganisation and women's work in Guadalajara, Mexico, *Bulletin of Latin American Research* **7**(2): 207–23.

—— (1991) Family well-being, food consumption and survival strategies during Mexico's economic crisis, in M. González de la Rocha and A. Escobar (eds) *Social Responses to Mexico's Economic Crisis of the 1980s*. Center for US–Mexican Studies, UCSD, San Diego: 115–127.

—— (1994) *The Resources of Poverty: Women and Survival in a Mexican City*. Basil Blackwell, Oxford.

—— (1995) Social restructuring in two Mexican cities: an analysis of domestic groups in Guadalajara and Monterrey, *European Journal of Development Research* **7**(2): 389–406.

—— (1997) The erosion of the survival model: urban responses to persistent poverty. Paper prepared for the UNROSD/UNDP/CDS Workshop 'Gender, poverty and well-being: indicators and strategies'. Trivandrum, Kerala, 24–27 November.

—— (1999) Hogares de jefatura femenina en México: patrones y formas de vida, in González de la Rocha, M. (ed.) *Divergencias del Modelo Tradicional: Hogares de Jefatura Femenina en América Latina*. Centro de Investigaciones y Estudios Superiores en Antropología Social, Mexico DF: 125–53.

González de la Rocha, M., Escobar, A., and **Martinez Castellanos, M. de la O.** (1990) Estrategias versis conflictos: reflexiones para el estudio del grupo doméstico en epoca del crisis, in de la Peña, G., Durán, J.M., Escobar, A. and García de Alba, J. (eds) *Crisis, Conflicto y Sobrevivencia: Estudios Sobre la Sociedad Urbana en México*. Universidad de Guadalajara/CIESAS, Guadalajara: 351–67.

Green, D. (1991) *Faces of Latin America*. Latin America Bureau, London.

—— (1995) *Silent Revolution: the Rise of Market Economics in Latin America*. Cassell in association with Latin America Bureau, London.

—— (1996) Latin America: neoliberal failure and the search for alternatives, *Third World Quarterly* **17**(1): 109–22.

Grosh, M. (1994) *Administering Targeted Social Programs in Latin America: From Platitudes to Practice*. World Bank, Washington, D.C.

Gudmundson, L. (1986) *Costa Rica Before Coffee: Society and Economy on the Eve of the Export Boom*. Louisiana State University Press, Baton Rouge.

Güendel, L. and **González, M.** (1998) Integration, human rights and social policy in the context of human poverty, in UNICEF (ed.) *Adolescence, Child Rights and Urban Poverty in Costa Rica*. UNICEF/HABITAT, San José: 17–31.

Gutmann, M. (1996) *The Meanings of Macho: Being a Man in Mexico City*. University of California Press, Berkeley.

—— (1997) The ethnographic (g)ambit: women and the negotiation of masculinity in Mexico City, *American Ethnologist* **24**(4): 833–55.

—— (1999) A manera de conclusión: solteras y hombres. Cambio e historia, in González de la Rocha, M. (ed.) *Divergencias del Modelo Tradicional: Hogares de Jefatura Femenina en América Latina*. Centro de Investigaciones y Estudios Superiores en Antropología Social, Mexico DF 163–72.

Instituto Nacional de la Mujer (INAMU) (1998) *Maternidad y Paternidad: Dos Caras del Embarazo Adolescente*. INAMU, San José.

Jayawardena, C. (1960) Marital instability in two Guianese sugar estate communities, *Social and Economic Studies* 9: 76–100.

Jelin, E. (1991) Introduction: everyday practices, family structures, social processes, in Jelin, E. (ed.) *Family, Household and Gender Relations in Latin America*. Kegan Paul International/UNESCO, London/Paris: 1–5.

Kaztman, R. (1992) Por qué los hombres son tan irresponsables?, *Revista de la CEPAL* 46: 1–9.

Lancaster, R. (1992) *Life is Hard: Machismo, Danger and the Intimacy of Power in Nicaragua*. University of California Press, Berkeley.

—— (1997) Guto's performance: notes on the transvestism of everyday life, in Balderston, D. and Guy, D. (eds) *Sex and Sexuality in Latin America*. New York University Press, New York: 9–32.

LeVine, S. (in collaboration with Clara Sunderland Correa) (1993) *Dolor y Alegría: Women and Social Change in Urban Mexico*. University of Wisconsin, Madison.

Lomnitz, L. and **Pérez-Lizaur, M.** (1991) Dynastic growth and survival strategies: the solidarity of Mexican grand-families, in Jelin, E. (ed.) *Family, Household and Gender Relations in Latin America*. Kegan Paul International/UNESCO, London/Paris: 123–32.

López, M. de la Paz and **Izazola, H.** (1995) *El Perfil Censal de los Hogares y las Familias en México*. Instituto Nacional de Estadística, Geografía e Informática, Aguascalientes.

Lumsden, I. (1996) *Machos, Maricones and Gays: Cuba and Homosexuality*. Temple University Press/Latin America Bureau, Philadelphia/London.

McClenaghan, S. (1997) Women, work and empowerment: romanticising the reality, in Dore, E. (ed.) *Gender Politics in Latin America: Debates in Theory and Practice*. Monthly Review Press, New York: 19–35.

McIlwaine, C. (1993) Gender, Ethnicity and the Local Labour Market in Limón, Costa Rica. Unpublished PhD thesis, Department of Geography, London School of Economics.

—— (1997) Vulnerable or poor? A study of ethnic and gender disadvantage among Afro-Caribbeans in Limón, Costa Rica, *European Journal of Development Research* 9(2): 35–61.

Ministerio de Planificación Nacional y Política Económica (MIDEPLAN) (1995) *Estadísticas Sociodemográficas y Económicas Desagregadas por Sexo, Costa Rica, 1980–1994.* MIDEPLAN, San José.

Miraftab, F. (1994) (Re)production at home: reconceptualising home and family, *Journal of Family Issues* **15**(3): 467–89.

Mirandé, A. (1997) *Hombres y Machos: Masculinity and Latino Culture.* Westview, Boulder, CO.

Molina, G. (1993) *Cómo Obtener Protección Legal?* Centro Feminista de Información y Acción, San José.

Molyneaux, M. (1996) *State, Gender and Institutional Change in Cuba's 'Special Period': the Federación de Mujeres Cubanas,* Research Paper 43. Institute of Latin American Studies, University of London.

Momsen, J. (1992) Gender selectivity in Caribbean migration, in Chant, S. (ed.) *Gender and Migration in Developing Countries.* Belhaven, London: 73–90.

Montoya Tellería, O. (1998) *Nadando Contra Corriente: Buscando Pistas para Prevenir la Violencia Masculina en las Relaciones de Pareja.* Puntos de Encuentro, Managua.

Moore, H. (1994) *Is There a Crisis in the Family?* Occasional Paper 3, World Summit for Social Development, UNRISD, Geneva.

Patterson, S. (1994) Women's survival strategies in urban areas: CARICOM and Guyana, in Meer, F. (ed.) *Poverty in the 1990s: the Responses of Urban Women.* UNESCO/International Social Science Council, Paris: 117–33.

Pearson, R. (1997) Renegotiating the reproductive bargain: gender analysis of economic transition in Cuba in the 1990s, *Development and Change* **28**: 671–705.

Peña Saint Martin, F. (1996) *Discriminación Laboral Femenina en la Industria del Vestido de Mérida, Yucatán,* Serie Antropología Social. Instituto Nacional de Antropología e Historia, Mexico DF.

Powell, D. (1986) Caribbean women and their response to familial experiences, *Social and Economic Studies* **35**(2): 83–130.

Prieur, A. (1996) Domination and desire: male homosexuality and the construction of masculinity in Mexico, in Melhuus, M. and Stølen, K.A. (eds) *Machos, Mistresses and Madonnas: Contesting the Power of Latin American Gender Imagery.* Verso, London: 83–107.

Proyecto Estado de la Nación (PEN) (1998) *Estado de la Nación en Desarrollo Humano Sostenible.* Proyecto Estado de la Nación, San José.

Proyecto Estado de la Región (PER) (1999) *Estado de la Región en Desarrollo Humano Sostenible.* Proyecto Estado de la Nación, San José.

Pulsipher, L. (1993) 'He won't let she stretch she foot': gender relations in traditional West Indian houseyards, in Katz, C. and Monk, J. (eds) *Full*

Circles: Geographies of Women over the Life Course. Routledge, London: 107–21.

Quiroga, J. (1997) Homosexualities in the tropic of revolution, in Balderston, D. and Guy, D. (eds) *Sex and Sexuality in Latin America*. New York University Press, New York: 133–51.

Radcliffe, S. (1999) Latina labour: restructuring of work and renegotiations of gender relations in contemporary Latin America, *Environment and Planning A* **31**: 196–208.

Robles Berlanga, H. with **Artís, G., Salazar, J.** and **Muñoz, L.** (2000) ¡ . . . Y *Ando Yo en el Campo! Presencia de la Mujer en el Agro Mexicano*. Procaduría Agraria, Mexico DF.

Rudolf, G. (1999) *Panama's Poor: Victims, Agents and Historymakers*. University Press of Florida, Gainesville.

Safa, H. (1995a) *The Myth of the Male Breadwinner: Women and Industrialisation in the Caribbean*. Westview, Boulder, CO.

—— (1995b) Economic restructuring and gender subordination, *Latin American Perspectives* **22**(2): 32–50.

—— (1999) *Women Coping With Crisis: Social Consequences of Export-Led Industrialisation in the Dominican Republic*, North–South Agenda Paper No 36. North–South Center, University of Miami, Miami.

Safilios-Rothschild, C. (1990) Socio-economic determinants of the outcomes of women's income-generation in developing countries, in Stichter, S. and Parpart, J. (eds) *Women, Employment and the Family in the International Division of Labour*. Macmillan, Basingstoke: 221–8.

Sarduy, P.P. and **Stubbs, J.** (1995) Introduction, in Minority Rights Group (ed.) *No Longer Invisible: Afro-Latin Americans Today*. Minority Rights Group, London: 1–8.

Selby, H., Murphy, A. and **Lorenzen, S.** (1990) *The Mexican Urban Household: Organizing for Self-Defence*. University of Texas Press, Austin.

Smith, R.T. (1988) *Kinship and Class in the West Indies: a Geneaological Study of Jamaica and Guyana*. Routledge and Kegan Paul, London.

Sørensen, N. (1985) Roots, routes and transnational attractions: Dominican migration, gender and cultural change, in Peek, P. and Standing, G. (eds) *State Policies and Migration: Studies in Latin America and the Caribbean*. London, Croom Helm: 173–205.

Steiner, H. and **Alston, P.** (1996) *International Human Rights in Context: Law, Politics and Morals*. Clarendon Press, Oxford.

Townsend, J., Zapata, E., Rowlands, J., Alberti, P. and **Mercado, M.** (1999) *Women and Power: Fighting Patriarchies and Poverty*. Zed, London.

Trotz, A. (1996) Gender, ethnicity and familial ideology in Georgetown, Guyana: household structure and female labour force participation reconsidered, *European Journal of Development Research* **8**(1): 177–99.

United Nations (UN) (1995) *The World's Women 1995: Trends and Statistics.* UN, New York.

—— (1997) *Demographic Yearbook 1995,* UN, New York.

United Nations Development Programme (UNDP) (1995) *Human Development Report 1995.* Oxford University Press, New York.

—— (1997) *Human Development Report 1997.* Oxford University Press, New York.

—— (1998) *Human Development Report 1998.* Oxford University Press, New York.

United Nations Economic Commission for Latin America and The Caribbean (UN/ECLAC) (1994) *Social Panorama of Latin America.* ECLAC, Santiago.

Valverde, J.M. (1998) Recent urbanisations and quality of life of adolescents at high social risk, in UNICEF (ed.) *Adolescence, Child Rights and Urban Poverty in Costa Rica.* UNICEF/HABITAT, San José: 49–64.

Vance, I. (1987) More than bricks and mortar: women's participation in self-help housing in Managua, Nicaragua, in Moser, C. and Peake, L. (eds) *Women, Human Settlements and Housing.* Tavistock, London: 139–65.

Vincenzi, A. (1991) *Código Civil y Código de la Familia.* Lehmann Editores, San José.

Ward, K. and **Pyle, J.** (1995) Gender, industrialization, transnational corporations and development: an overview of trends and patterns, in Bose, C. and Acosta-Belén, E. (eds) *Women in the Latin American Development Process.* Temple University Press, Philadelphia: 37–64.

Williams, C. (1986) The role of women in Caribbean culture, in Ellis, P. (ed.) *Women of the Caribbean.* Zed, London: 109–14.

Willis, K. (1993) Women's work and social network use in Oaxaca City, Mexico, *Bulletin of Latin American Research* **12**(1): 65–82.

—— (2000) No es fácil pero es posible: the maintenance of middle class women-headed households in Mexico, *European Review of Latin American and Caribbean Studies* 69: 29–45.

World Bank (1995) *World Development Report 1995.* Oxford University Press, New York.

—— (1996) *World Development Report 1996.* Oxford University Press, New York.

Yudelman, S. (1989) Access and opportunity for women in Central America: a challenge for peace, in Ascher, W. and Hubbard, A. (eds) *Central American Recovery and Development: Task Force Report to the International Commission for Central American Recovery and Development.* Duke University Press, Durham: 235–56.

CHAPTER 3

Ethnicities, nationalism and racism

David Howard

Introduction

> 'Race and often the biologized conceptions of ethnicity have been potent factors in the making of the Americas.'
>
> (Oostindie 1996: 1)

> 'Because colour, class, and personal identity are all intimately linked, racial awareness suffuses every aspect of West Indian life.'
>
> (Lowenthal 1972: 320)

This chapter seeks to illustrate the influences of race and ethnicity in Caribbean and Central American societies, noting the subtle complexities of ethnic identity across a varied range of territories, while allowing an overview of key aspects. The region has experienced comparable histories of European colonial influence, the widespread suppression of indigenous peoples and the brutal legacy of slavery from the beginning of the sixteenth century. Insularity, the dominance of the sugar plantation system and African slave labour initially separated the Caribbean from the mainland experience. Thus, the national identities of the Caribbean have been largely Creole-dominated, arising from European and African influences, whereas those of the mainland are predominantly Spanish and indigenous.

Indigenous and non-indigenous peoples in the region have been classified along social lines, rather than by physical appearance alone. Under Spanish colonial rule, the indigenous inhabitants were those who paid taxes or tribute, whereas the *mestizos/as* spoke Spanish, did not live in designated indigenous settlements, and generally were not forced to pay the same level of fiscal dues (Wearne 1994: 6). *Mestizaje*, the concept of a 'putative joint racial and cultural descent . . . derived from both the Spaniards and the Indians after the conquest', has obscured the reality of racism

---

Box 3.1   Definitions of ethnic classification terms

This chapter highlights the diversity of terms, based on both physical appearance and cultural practices, that are used in the region. The following list provides basic definitions of the most widely used terms. In Spanish, many words end in 'o' when referring to men and 'a' when the subject is a woman.

Creole: Caribbean term referring to the mixture between European and African influences.

Criollo: Spaniard born in the Americas.

Indígena: An indigenous person.

Indio/a: Literal meaning is 'Indian', but usually used in a derogatory way to refer to indigenous populations in Mexico and Central America.

Ladino/a: Person of shared European and indigenous descent in Latin America. Often used in relation to adoption of 'western' styles of dress, land-holding and urban living.

Mestizaje: The concept of a mixing of European and indigenous racial and cultural characteristics in Latin America.

Mestizo/a: Person of shared European and indigenous descent in Latin America.

Moreno/a: Literal Spanish meaning is 'brown person'. Caribbean term for a person of mixed African and European ancestry.

Mulato/a: Caribbean term for a person of mixed African and European ancestry.

---

and discrimination against indigenous peoples in much of the region (Gutiérrez 1998: 4). Miscegenation has characterised the demographic development and evolution of racialised identities in Middle America. The term *mestizo/a* historically represented elements of continuity with Spanish culture, incorporating a polarizing identity that distanced Catholic, bilingual and 'Hispanic' peoples from the indigenous communities (Knight 1990: 74). The subjectivity of the *mestizo/a*-indigenous continuum clearly invalidates strict differentiation by aesthetic and social categories, nevertheless *mestizaje* remains an important aspect of national and ethnic consciousness throughout Central America (Esteva-Fabregat 1995). Vasconcelos' (1920) concept of the 'cosmic race' embodied the spiritual and creative fusion of European and indigenous ancestry, which shrouded the perceived deficiencies of the latter in the former's cloak of supposed superiority. Box 3.1 outlines some commonly used terms relating to ethnic groupings.

As the colonial tide receded on the mainland during the nineteenth century, the term *mestizo/a* became prominent as national identities were forged in the fledgling republics. In Nicaragua, for example, the failure to emphasise the size and influence of the indigenous population in favour of a *mestizo/a* majority led to the 'myth of *mestizaje*' (Gould 1998). Indigenous peoples were marginalised economically, socially and rhetorically. The label *indio/a* itself continues to be uttered in disparaging terms by prejudiced individuals, who associate it solely with low social and economic status. The denial of indigenous expression, however, at times may strengthen an identity in the form of collective memory and shared resistance against discrimination: 'Ask an *indígena* in Guatemala to define themselves today and they will most probably do so by village or community first, linguistic group second and only by western perceptions of nationality, i.e. Guatemalan third . . . they see themselves belonging to a community that shares basic cultural values by which it differentiates itself from the rest of society' (Wearne 1994: 6).

Similar experiences of colonial domination and resistance have shaped the societies of the Caribbean. The legacy of European colonial powers and subsequent social structure, economic relations of dependency, international migration and multiculturalism cast up a myriad of images and factors that shape what it is to be Caribbean (Edmondson 1999). The region consists of a medley of ethnic identities and strong national rivalries that exist between the territories and island states, which in many instances negate a viable regional identity. A federation of ten countries within the former British Caribbean lasted only two years before inter-island disputes led to its demise in 1960. Benítez-Rojo (1992) has attempted to unravel the apparent confusion of island identities, and recounts their ordered repetition throughout the Caribbean islands, most notably via the framework of plantation society. The plantation, he argues, is the commonality from which to explore the region's ethnicities.

Perhaps cricket, suggests Beckles (1999), creates the forum in which former colonies, the English-speaking West Indies, contest relations of power and nation with their one-time coloniser. The former Spanish, Dutch and French colonies provide quite a different context of what it is to be Caribbean. Among the more populous and larger territories of Cuba, the Dominican Republic and Puerto Rico, the Caribbean would more aptly be represented as a baseball-playing, Spanish-speaking majority. Paradoxically, the very diversity of the islands is arguably the underpinning constant that unites the Caribbean as an imagined identity, beyond the aesthetic superficiality of sun, sand and sea.

Regional patterns and ethnic backgrounds

A symbolic characterisation of the Americas that divides it into three regions, namely Indo-America, Afro-America and Ibero-America, provides a rough basis for the mapping of dominant ethnic influences in Middle America (Wagley 1968). The predominance of European cultural attachment in Costa Rica includes its classification as part of Ibero-America along with much of South America, whereas the greater influence of indigenous populations throughout the rest of Central America and Mexico leads to their inclusion under Indo-America. The plantation economies of Caribbean territories are grouped as Afro-America, since these societies were dependent on the enslavement of African workers, unlike the mainland estates that in general utilised indigenous labourers as effective wards of the state. A similar categorisation of Middle America incorporates a twofold division between the Euro-African Caribbean Rimland and the Euro-Indian Mainland (Augelli 1989).

The singularity of such classifications is evidently false, yet provides an initial, albeit hazy overview from which we can focus more clearly on the complexities of ethnicity in the region. The populations of southern Mexico and Guatemala, for example, have a greater indigenous component, as opposed to the predominance of *mestizo/as* elsewhere in Central America. Unlike the Caribbean islands, mainland societies were characterised by the later development of plantation economies that relied on both *mestizo/a* and indigenous labour. The former Spanish colonies in the Caribbean – Cuba, Puerto Rico and the Dominican Republic – similarly experienced a delayed evolution of the plantation system and subsequently relied less extensively on enslaved African labourers compared with the British, Dutch or French territories. The popular legacy and impact of slavery has therefore been marginalised to some extent in the Hispanic Caribbean, until more recent revisions have assisted a progressive reappraisal of ethnic and national consciousness.

A fourfold typology of Caribbean societies initially serves to unravel the dominant but varied influences of racism, colonialism, slavery and the plantation system (Clarke 1991). Ethnic and social cleavages are highlighted by adopting frameworks of stratified or segmented pluralism, class stratification and folk communities. Plural-stratified societies include Jamaica, the Leeward Islands and the French and Dutch Antilles in which the major socio-economic divisions are deemed to correspond with ethnic affiliation. During slavery, the racialised classifications of white, mulatto and black corresponded to the legalised strata of citizens, freed former slaves and enslaved labourers. Haiti, French Guiana and the Windward Islands are also included in this category, yet differ in that the white upper stratum has declined, leaving the generalised separation between lighter, wealthier elites and the less affluent population, usually of African descent.

Plural-segmented societies encompass the broadly differentiated Creole-East Indian sectors of the Trinidadian, Guyanese and Surinamese populations and the Creole-*mestizo/a* divisions in Belize. The ethnically derived vertical cleavages in the Creole-East Indian societies are the result of the incorporation of indentured labour, largely from Asia, following the abolition of slavery in the mid-nineteenth century. Hispanic territories in the Caribbean are categorised as class-stratified societies since class or socio-economic stratification is considered to be dominant over ethnic division. Cuba, Puerto Rico and the Dominican Republic, as already mentioned, developed under the influence of Spanish colonialism and on a different timescale from the rest of the region. These countries experienced two main periods of expanding sugar production, during the sixteenth and nineteenth centuries. The development of sugar plantation economies during the nineteenth century and the relatively higher incidences of white immigration or conditions of free labour meant that these countries evolved as societies with slaves rather than slave societies (Clarke 1991: 9). Finally, folk societies are those largely undifferentiated populations of the smallest islands that have traditionally relied on peasant agriculture and subsistence activities.

In the UK, there is a tendency to equate Caribbean characteristics solely with those of the British Commonwealth, but this characterisation of Caribbean societies has been roundly contested (Hoetink 1985). Hispanic societies constitute two-thirds of the Caribbean population, and account for three-quarters of the land surface area. The importance of linguistic homogeneity and the influence of Catholicism as unifying factors among the former Spanish territories formed a crucial backbone for Hispanic nation-building. For almost two centuries after 1492, the entire Caribbean region was part of the Spanish colonial empire. The goals of capital accumulation promoted the process of colonisation and concomitant incorporation of slavery into the acquisition of territory and European settlement (Williams 1944). Spain neglected the region after its initial arrival to pay more attention after 1519 to the mainland with its perceived greater opportunities. Only towards the end of the eighteenth century did the Spanish colonial powers seek to renew their interests in the territories of Cuba and Puerto Rico, resuming the slave-based plantation economies and metropolitan dominance of island affairs.

Ethnic and racial affiliations have been and remain central tenets in the formation and organisation of Middle American states and nationalities. The diverse indigenous societies of Middle America suffered inevitable transformation following the initial European intrusion and enforced colonisation at the end of the fifteenth century. Social antagonisms and the economic and political imperatives of colonial expansion underpinned the development of post-Columbian territorial organisation, led first by Spanish exploration and subsequently by the imperial powers of England, France and Holland, and to a lesser extent Denmark and Sweden. In the twentieth century, the US for the most part, and later the former Soviet Union, entered the fray. Colonisation and the metropolitan drive to secure political, economic and cultural dominance thus created spheres of influence, which have shaped the evolution of regional and national alliances and shared identities.

Five centuries of colonial and external influence in Middle America, successive migrations, the imposition of slave-based societies in the Caribbean and the rapid development of a mainland *mestizo/a* population since the seventeenth century have created a regional diversity that largely defies attempts at categorisation into exclusive groups. Ethnic labelling is very much dependent on the perspective of description or ascription: 'Although there is no universally accepted definition of indigenous peoples, it is worth remembering that, like the term "Indian", it is an externally imposed and relative term. The indigenous peoples of the Americas are only indigenous because they and their territories were colonised by others' (Wearne 1996: 11). Similarly, the fluidity of racial labelling on the basis of aesthetics has obvious problems, particularly when notions of status are closely correlated with racial or ethnic terminology: 'The problem is one exacerbated by the incredible variety of terms used to describe people whose appearance might be quite similar: Black, Afro-American, Creole, Mulatto, Garífuna, Black-Indian, Black Carib, Antillean Negro, Moreno, *pardo, negre, preto* – the list is extensive' (Purcell 1993: 9). In general, elite and popular discourses in Middle America promote a bias that gives higher status to European or white ancestry and appearance, and negates indigenous and African ethnicities.

Indigenous peoples suffered severely during the process of military conquest by the Spanish, by the sword or through contact with endemic European disease. Remarkably, by the end of the seventeenth century the indigenous population on the mainland had re-established itself, but in the new demographic and social context of European settlement and enforced labour. There are now some 40 million indigenous peoples in the Americas, 90 per cent of whom live in Mexico, Guatemala, Ecuador, Bolivia and Peru, making up 6 per cent of the total population. In Guatemala, 5.4 million indigenous citizens represent the majority, 60.1 per cent of the total population (Warren 1998: 8). Mexico has the largest number of indigenous peoples, over 10.5 million, although they make up only 12.4 per cent of the national population. The significant economic and social inequalities experienced by these populations is emphasised spatially, since the majority are located in southern Mexico in 13 states officially designated as 'eminently indigenous' (Barry 1992: 223). Unlike the mainland communities, there are few surviving indigenous peoples on the Caribbean islands, consisting largely of the one thousand residents on the island of Dominica. The subjectivity of classifying ethnicities is evident in the case of El Salvador, where official statistics designate the indigenous population to be 190,000, compared with an estimated 500,000 people who identify themselves as indigenous or *naturales*.

The Maya are the largest collective of indigenous peoples, numbering over 9 million, speaking over 20 different but related languages. They are concentrated in southern Mexico, Guatemala, Belize and Honduras. The forging of cross-class alliances on the basis of the Maya languages as communities of shared linguistic and cultural practices has promoted a pan-Mayan identity (see Box 3.2). The next most numerous indigenous group are the one million Nahuatl-speakers in Central Mexico and Mexico City.

Indigenous groups of Middle America have suffered overt discrimination since the beginning of the colonial period. Perhaps the most savage recent example of hate-driven racist violence against indigenous groups occurred in Guatemala during the early 1980s. Between 1982 and 1986 tens of thousands of people in the highlands were killed indiscriminately by the armed forces who assaulted rural settlements in an attempt to root out guerrilla forces and those sympathetic to their cause (REHMI 1999). Thirty years previously, a military coup had brought the downfall of a democratic socialist government which had pursued land reform policies in order to redress the gross social and economic inequalities in Guatemalan society. The 1954 rebellion pitted urban *ladinos/as* against rural Maya peoples, roughly corresponding to the affluent oligarchy and impoverished masses respectively. Three decades of state-led terror left one million people uprooted from their homes. During the 1980s alone, government repression against rural-based guerrilla movements targeted indigenous groups and led to 100,000 refugees being displaced from the Huehuetenango, San Marcos and El Quiché regions. Many fled to Mexico, to urban areas within Guatemala, or faced relocation in government-controlled 'model villages', while others sought refuge in the jungle and established Communities of

---

Box 3.2   Guatemala: Academy of Mayan Language

Mayan peoples form the largest indigenous group in Middle America, numbering approximately 5 million. The majority live in Guatemala, but also in southern Mexico, Belize and western areas of El Salvador and Honduras. Following the abatement of government-sponsored violence and repression during the early 1980s in Guatemala, 1987 marked a fresh beginning and cycle in the Mayan calendar. The so-called 'new dawn' was the spur for a rein-vigorated pan-Mayan identity, adopted by the indigenous rights movement, which brought together over 21 ethnic and related linguistic groups and sought to challenge collectively the ongoing discrimination and derogative treatment of the indigenous communities. The Mayan language was a key focus for unity and led to the foundation in 1990 of the Academy of Mayan Language to support indigenous cultures and promote bilingual teaching in Mayan ar-eas. A peace accord on the Identity and Rights of Indigenous People has provided further constitutional backing for the recognition of a multi-ethnic state since 1995. The following year, a multilateral agreement between Belize, Guatemala, El Salvador and Honduras to co-operate in the conservation and promotion of Mayan archaeological sites was ratified, which will provide further international and institutional support for the Mayan revival.

Source: O'Kane (1999).

Popular Resistance (O'Kane 1999). The strength of indigenous resistance remains a testament to the enduring sentiment of ethnic belonging.

Ethnicity, gender and race: a theoretical overview

The changing experiences of people's everyday lives and situations mean that the perceptions of their own identities may vary continuously. To understand any real sense of identity, the local specificities of place and context need to be known. This is especially true in Middle America, where slight differences in linguistic ability, physical appearance or even categories of friendship may be important indicators of ethnic allegiance (Mintz 1974: 320). The following section provides a theoretical outline to some of these subtleties of ethnicity, race and gender.

An ethnic group relates to a collectivity of people who are conscious of having common origins, interests and shared experiences, while race may more specifically be defined as a means of discriminating against social groups on the basis of their biological or cultural differences (Anthias 1990; Jackson and Penrose 1993). Given the close relationship between the two terms, however, to separate them via rigid definition is to miss the point of their inter-relation. Racist discourse clearly incorporates ethnic categories (which might be constructed around cultural, linguistic, territorial or supposed biological differences). As such, racial discrimination can be defined as the active expression of racism which aims to deny members of certain ethnic groups equal access to limited and valued resources (Cashmore 1995: 273).

Racism has frequently been defined with respect to the dual notion of winners and losers, the fortunate and the ill-fated. It has become increasingly 'naturalised' and disguised in discussions of alleged cultural and national norms. The distinction made by Hall (1990) between overt and inferential racism is poignant for the Middle American context. The latter form of racism is expressed by nationalistic terms of difference between peoples, incorporating expressions that refer to allegedly natural racial divisions, but inevitably are placed in a prejudicial framework. The former is more openly racist, emphasising racial categorisation above the individual. The Haitian population in the Dominican Republic, for example, is discriminated against on the grounds of different ethnic ancestry, nationality and cultural traits. Dominican antagonism towards Haiti is, thus, legitimised as a form of patriotism (see Box 3.3).

The potency of prejudice lies in its embeddedness as a perceived natural phenomenon, often 'learnt' during everyday interactions. Theories of socialisation propose that social and cultural attributes are formed principally during childhood, hence the importance of the household environment. Racial and ethnic identities and values are expressed within the domestic sphere and come to form part of the socialisation process. The household, a unit of domestic reproduction and shelter, acts as 'a conduit for wider familial and gender ideologies', providing the context for the reproduction of patriarchal and racist culture (Chant 1996: 12). Two contradictory currents interact. First, the domestic sphere is the location for the formation and reproduction of race and gender relations. Second, and conversely, the household is also a site of resistance to racism and patriarchy, the latter defined as 'a fluid and shifting set of social relations in which men oppress women, in which different men exercise varying degrees of power and control, and in which women resist in diverse ways' (Hondagneu-Sotelo 1992: 393).

Race and gender relations are mutually constitutive, each operating within or around the other (Jackson 1994). Thus, any concept of race will be gendered, and any notion of gender racialised. First, racial variation differentiates the experiences of women *vis-à-vis* those of men in the labour market; secondly, men and women of similar race or ethnicity will be exposed to differing forms of gender discrimination (see Chant, this volume). Consequently, racism and patriarchy are not constant influences but vary according to time, place and context.

A popular children's song in the Spanish-speaking Caribbean presents an image of women which reproduces in verse the social structures of patriarchy. The lyrics concern a man setting out his requirements for a woman whom he would marry. He suggests that the ideal wife should be a wealthy widow, skilled at sewing and cooking. She must have all the necessary domestic skills to fulfil the role of a 'good housewife'. The repetition of such verses indoctrinates and reproduces patriarchal structures during the formative years of social development. It encapsulates the

Box 3.3 'The Haitian problem'

Popular culture juxtaposes an allegedly white, Hispanic and Catholic culture of the Dominican Republic with the African ancestry, *vodú* and presumed barbarity of its Haitian neighbour. 'The Haitian problem' arises from the on-going antagonism between the two countries, sustained by two centuries of history, and by the increasing numbers of Haitian labourers in the Dominican Republic. The manner in which 'Dominicanness' has been portrayed in relation to Haiti has coloured, or perhaps more accurately bleached, the image of the Dominican nation. Two processes have operated historically at popular and governmental levels – firstly, one of *blanqueamiento* or whitening, either by encouraging European immigration, for example during the 1930s, or by maintaining a social and cultural white bias, secondly, the official and popular propagation of the Haitian population as a threat to the Dominican nation.

Given their shared border, there has been a long history of migration between the two countries. Haitian migration to the Dominican Republic grew in influence with the development of the modern sugar industry from the 1870s. The Haitian presence in the Dominican Republic increased noticeably during the early twentieth century as a result of contract work on the sugar plantations. Following the sharp fall in sugar prices in the 1920s, Haitian labourers began to replace the labour force on the sugar plantations which consisted to a large extent of migrant workers from British territories in the Caribbean. The Haitians worked for much lower wage rates.

Estimates vary for the number of Haitians living in the Dominican Republic. Some suggest up to 1.5 million, but there are probably around 500,000 Haitians and Dominicans of Haitian descent living on Dominican territory. The size of the population of Haitian origin in Dominican Republic is said to have doubled in past ten to fifteen years. The Haitian sugar workers live mostly in rural communes, called *bateyes*, under conditions that have been equated with slavery by international human rights organisations. A quota system in which the Dominican government paid the Haitian authorities for each Haitian labourer existed up until 1986. It continues to operate today, albeit informally or via agreement and payment between the countries' military forces. More noticeable are the regular violent round-ups and deportations of Haitian workers by the Dominican military. These break up families and exile thousands of documented Haitians and Dominican labourers alike.

Source: Howard (1998).

conventional and widespread conception that domestic matters are women's issues (Walby 1990).

Darker-skinned women in the context of a highly racialised and patriarchal society would predictably suffer twice from racist and sexist subordination. Sexuality incorporates the aesthetics of the body through the lens of patriarchal relations and racial perceptions. The body is politicised, sexually objectified with respect to the established political and social force of heterosexuality. A book of well-known stories from the Hispanic Caribbean illustrates a common notion of female beauty – blond hair, blue eyes and white skin. Conceptions of race are entwined with sexuality and the aesthetic evaluation of beauty. The lighter the skin colour, the straighter the hair or the narrower the facial features, the greater the tendency to define beauty within these terms. A Dominican television channel had two dark-skinned presenters for an evening programme. The station soon became known in popular conversation as the 'Channel of the Uglies'.

Employment advertisements seek potential employees of *buena presencia* or who are *bien parecida* or *aparente* (all meaning 'nice looking'). These solicitations for 'a good appearance' conform to the light colour bias. Sexually and racially biased advertising, often overt, is commonly used in the media. A brand of Dominican rum has an established marketing campaign featuring light-skinned models with the slogan, *una cosa de hombres* – the bottle of rum and the woman are 'men's things'. The broadcast times for a new television commercial for the brand of rum were advertised in newspapers days before the commercials were first aired during the summer of 1994; the Dominican male audience eagerly awaited the adverts. The light-skinned model was clothed in a transparent white shirt, playing a saxophone on a beach. The newspaper advertisements attempted to tease the reader with a photo of the model: 'You have to see it! It's sensual, it's daring . . . it's a man's thing.'

In similarly stark terms, a widely sold tonic for improving male sexual potency, *Fórmula Árabe*, reinforces the sexual, and racial, message of its market strategy. The packaging of the tonic has a photo of a muscular, bearded and tanned man squeezed into a pair of jeans, sporting a turban and standing in a sexually assertive pose. Between his straddled legs, clinging on to his thigh, and gazing upwards, sits a white model with long blond hair in a white silk dress. Race and sexuality sell products, but more specifically it is the promised image of a submissive woman's body, white or light-skinned, which sells.

The white bias in beauty challenges the desirability of black sexuality. The contradiction is clear – popular male and female opinion reveres light-skinned beauty, yet popular sexual myth imbues the *negro/a* or *mulata/o* with vigour and skilful prowess. An informative comparison may be made with apartheid South Africa, where interwoven notions of sexuality and race reinforced the aesthetics of prejudice.[1] Bastide (1961: 18) outlines the sexual stereotypes that surround racial classification: 'the Dusky Venus hides the debasement of the black woman as a prostitute; and the Black Apollo is seeking revenge on the white man. It is not so much that love breaks down barriers and unites human beings as that racial ideologies extend their conflicts even into love.' Similarly, Fanon (1986) views the desire of dark-skinned women for lighter men as an attempt to obtain whiteness, to gain legitimacy and bear children in the 'white world'. The craving for revenge, Fanon relates, drives darker-skinned men's desire for white women, and underpins the pathological embeddedness of racism in colonial and postcolonial societies.

[1] David Simon, pers. comm.

The interaction of race and gender has implicit importance for concepts of nationalism: 'As reproducers of the next generation of national citizens, women have been viewed as crucial boundary-markers in gender nationhood . . . Given the emergence of modern nation states out of colonial, "non-national" encounters, the issue of race is central to questions of belonging, boundary-marking and identity' (Radcliffe 1999: 215). Lancaster's (1992) analysis of Nicaraguan nationality traces the fields of power invested in racial and sexual hierarchies to illustrate how the Sandinista revolutionary triumph in 1979 served to limit homosexual expression through its doctrine of public order and *machista* morality. *Machismo* emphasises male virility and dominance, concomitant with submissive notions of womanhood. Nevertheless, Lancaster accounts for the presidential victory of Violeta Chamorro over the Sandinista leader Ortega in the 1991 elections in terms of gendered persona: 'Chamorro appeared to be a beleaguered, sympathetic mother and grandmother, while Ortega appeared *machista* in a society grown tired and weary of *machismo*' (1992: 292).

Compatible development

Clearly ethnicity and gender are key influences for an understanding of Middle American societies, acting as simultaneous and interactive factors in social relations and affecting the ways in which people gain access to resources and participate in society. With this in mind, these relationships and interconnections have implicit importance for the provision of intra-regional development policy, which 'must not only be more "gender-aware" but also and in conjunction, more ethnically aware' (McIlwaine 1995: 241; see also McIlwaine, this volume).

These consequences are firmly incorporated in the concept of ethno-development, which provides that all economic and social development should be shaped by the requirements of the people, 'compatible with the cultural specificity and needs of ethnic groups' (Hettne 1995: 203). The ethnic heterogeneity of Middle American societies makes such a focus of key importance, as various groups gain greater access to governance and economic development initiatives. Ethnodevelopment policies provide minority voices with a wider platform on which to base specific needs and rights. Above all, such strategies rely on self-determination, the collective capacity for an ethnic group to control its destiny, involving negotiation and compromise with the state power. The rebellion of the Miskito communities in the Caribbean coastal region of Nicaragua led to the decree of the Autonomy Law in 1987, which legally established the rights of Atlantic Coast communities, and the subsequent creation of regional assemblies.

As a reaction against the intrusive state-led politics of the socialist Sandinista government in Nicaragua during the 1980s, the Miskito rebellion was a struggle for self-determination and autonomy among the indigenous peoples of the coast. The ethnic militancy of the Miskito groups may be explained in terms of an Anglo-affinity, stemming from a long-term association in the region with Anglo-American cultural institutions and practices, namely contacts with the Moravian church, North American companies and US Marines (Hale 1994: 200). The Miskitos generally held the Sandinista government to be 'the antithesis of every important premise that Anglo affinity entailed: Communist, atheist, against "economic freedom", hostile to the material aid that outsiders, especially Americans, always had provided' (Hale 1994: 202). Calls for autonomy and resistance to the Sandinista revolution forced the Miskito communities to accept the nationalist design of the revolution, or remain marginalised. The outcome of Miskito resistance, however, led to the establishment of the Northern and Southern Regional Autonomous Assemblies. The Autonomy Law recognised the right of the coastal communities 'to preserve their cultural identities, art and culture, as well as the right to use and enjoy their own waters, forests and communal lands for their own benefit' (Thornberry 1991: 21). Low turnout for elections in 1998 suggested that despite initial successes, there remains a lack of clarity concerning the exact powers of the regional assemblies, and more importantly, an ongoing intransigence on the part of the centralist government to divest authority.

The demand for self-determination, while not necessarily implying outright independence, recognises the failure of the monolithic nation-state to cater for the ethnic and cultural diversity within its boundaries (Hettne 1995). Strategists have learnt that they can no longer ignore religion, kinship and cultural practice in their thoughts on development theory (Worsley 2000). As such, a wider governmental awareness of the issues surrounding ethnodevelopment, accompanied by sensitive policy provision, is a fundamental concept to be promoted in a region as ethnically diverse as Middle America.

Nation-building and independence

This section addresses some of the complexities surrounding ethnicity and nationalism in the region, given the variety of historical and contemporary contexts. Nationalities are in a constant process of construction as histories progress and circumstances are modified. Whereas statehood and legislative independence may be sculpted through political leverage and conflict, nationalities are forged through the contested terrain of ethnic, racial and gendered ideologies.

In the regional context, Caribbean societies underwent a much later period of decolonisation compared to the majority of mainland states, which gained independence at various stages during the nineteenth century. Thus the Caribbean, while consisting of over 50 self-governing and independent states, is still partly shielded, or smothered by the fading blanket of European colonialism. In sharp contrast, the mainland states experienced virulent and often violent campaigns for independence.

Mestizo rules

Mexican independence in 1810, and the subsequent revolutionary years between 1910 and 1920, gave rise to new nationalist rhetoric that downplayed ethnic difference in favour of a unified fledgling state. Indigenous peoples, however, were still deemed to be inferior by the elite who were constructing the path along which the nation would evolve. During the final quarter of the nineteenth century, *mestizaje* was developed as the discourse for indigenous integration. Following the decade of revolution there was an increasing need to forge a post-revolutionary ideology for the new state. Strong civic and political concepts of nationhood directed the official ideology to emphasise ethnic and cultural unity.

Eurocentric constructions of *mestizaje* incorporate a whitening bias, yet have historically excluded *mestizo/as* from the largely white power structure. *Mestizo/a* thus becomes a legitimating ideal or subordinate representation, asserting notions of white, indigenous and black hierarchies of purity. Recognition of the limitations of *mestizaje* in recent years, however, has finally pushed aside notions of the cosmic race and racial lineage. An amendment to the Mexican constitution in 1991 emphasised an ideology of multiculturalism, the official endorsement of a multi-ethnic state which rejected notions of cultural conformity and integration into a single ethnic-ally based nationalist project. Today, the historical polarisation between *mestizos/as* and indigenous sectors of society continues to underpin profound social inequalities and racist ideologies (Gall 1998: 153). The uprising of the indigenous-based Zapatista Army for National Liberation in Chiapas in 1994 received international recognition for its challenge to traditional *mestizo/a* discourse and emphasis on local autonomy and respect for indi-genous cultures and languages (Morris 1999) (see also Box 3.4).

Box 3.4   The Chiapas rebellion

The first day of 1994 brought news of a well-organised popular rebellion in the southern Mexican state of Chiapas. The revolt was led by the Zapatista Army for National Liberation (EZLN), named after Emiliano Zapata's famous revolutionaries of the 1910s, and emphasised the chronic lack of assistance to indigenous and marginalised communities from the centralised Mexican government. The capture of several towns coincided with the activation of the North American Free Trade Agreement (NAFTA), which rebel leaders argued would only serve to worsen the poverty and social deprivation of the region's largely indigenous population. A heavy-handed response by the military sought to extinguish the rebellion, but only served to strengthen its cause as the enigmatic leader, Subcommander Marcos continued to offer resistance from the remote mountains, keeping in contact with the world's press and political organisations through regular and detailed email communiqués. After several months, the guerrillas and government entered protracted negotiations, but the problems of unequal land rights and ownership, limited popular access to centralised power or investment opportunities, and ongoing cases of govern-ment corruption and human rights abuses continue to marginalise the region's population.

Source: Russell (1995).

Concepts of *mestizaje* have also caused tensions within Belize, which only gained independence from Britain in 1981. As a 'West Indian enclave in an otherwise Hispanic isthmus . . . Belizean national identity thus reflects the ideology of nationhood in many plural societies, in which one culture is held to represent national identity and all others are politically marginalised or stigmatised as "liabilities" to the nation' (Moberg 1997: xxx). The emergence of *mestizos/as* as the largest ethnic group, challenging creole dominance, emphasised longstanding uncertainties concerning Belizean national identity. Since the arrival of Mexican war refugees in the late 1840s, Belize has had a substantial *mestizo/a* population. In addition, Guatemala has claimed sovereignty over the territory since the mid-nineteenth century. Spanish-speakers have therefore been regarded with suspicion by many Creoles in Belize, increasingly so since the 1991 census revealed that Spanish had become the most popularly spoken language. The departure of the British army garrison in 1993, followed a year later by renewed Guatemalan claims of territorial sovereignty, have served to heighten creole anxieties and rejection of *mestizaje*.

Officially, Belizean nationalism is multi-ethnic, a 'haven of cultural pluralism', with the two largest categories as Creole and Hispanic, then Garífuna (African and indigenous heritage) and Maya (Kroshus Medina 1997: 757). A pan-African Belizean identity, however, is increasingly apparent among Creoles and Garífuna who perceive Spanish-speakers as mutual adversaries: 'Immigration (from other Central and South American states) emerged as the most critical issue facing Belize by the late 1980s, when the country's population claimed the highest proportion of foreign-to native-born residents in the hemisphere' (Moberg 1997: xxix). Belizean 'nationality' remains a hotly contested identity.

Black nationalism and Creole consciousness

Insularity is a key aspect of Caribbean nationalism, since the locus of power has often been beyond domestic shores. External threat is often an important mobilising factor for the fruition of nationalism. Mintz argues that the region 'is as homogeneous as it is because of the twin forces of imperial imposition and popular response; it is differentiated as it is because that imposition was multiplex, diffuse, and conflictual, and because each responding population differed in significant degree from every other' (1974: 256). Lowenthal (1972), conversely, has stressed that the denial of West Indian identity is a consequence of slavery and a colonial legacy that denies ethnic and national self-assertion.

Modern Caribbean states were founded under a colonial gaze that chose to juxtapose the alleged backwardness of indigenous and African histories with the progress of European civilisation. During the reworking of colonialism during the 1940s up to independence in the 1960s, new political spaces were opened up for the African Creole majority to participate in governmental and regional organisations during the process of decolonisation. Access to power entailed certain privileges that the African Creole elite were loath to share with co-resident ethnic groups.

Haiti, however, stands out as a unique case of rebellion and subsequent statehood formed on the premise of anti-white insurrection and subsequent black nationalism. Haitian independence was declared in 1804, liberation arising from a 12-year revolt by slaves of African descent. The struggle had been fought on the basis of a tenuous alliance between two groups of former slaves (Nicholls 1996: 37). On the one hand, there were those who had gained their freedom on the basis of birthright, as the offspring of the white planter class, and who were largely *mulâtre* (mulatto) property owners. On the other, there were the black plantation workers who were more recently liberated by the abolition of slavery in 1793. As a result, colour distinction became the basis for class or status formation in the new republic. The *mulâtre* population was representative of neither the black nor white populations; they enjoyed the benefit of social and economic privileges yet did not have the status of the former white colonists. The *mulâtre* elite retained sole access to private property following independence, whereas their increasingly bitter allies, the former slaves, remained without land. Haiti's first ruler, Jean-Jacques Dessalines, moved to confront disparities between the two groups, dividing estates among the freed labourers, while the first constitution declared that all Haitians were to be called black, regardless of skin colour. Today, Haitians are proud of their historical status as the first black republic in the Western Hemisphere, but the economic disparities, embedded along racial lines, remain as harsh and severe as ever.

A more contemporary and perhaps globally recognisable proponent of Black Power revolutionised expectations not only in Middle America, but with dramatic impact in North America and Europe. Marcus Garvey, who emigrated from Jamaica to the US, perhaps delivered the most vocal programme in support of pan-African awareness during the 1920s. He travelled throughout Central America and the Caribbean, establishing publications that campaigned for decolonisation and African liberation (Davis 1995). Key to his plans for African Creole self-development and deliverance from white colonial oppression was the return to Africa. He founded the Universal Negro Improvement Association in the US and the Black Star Steamship Line to assist the organisation of the return passage to the African continent, believing that repatriation to the Promised Land of Ethiopia would lead to the redemption of all Africans exiled in the world of white oppression. The religious dimension of Garvey's ideas has its framework in Rastafarianism, which heralds King Haile Selassie of Ethiopia, crowned in 1930, as the Black Messiah and direct descendent of King Solomon. Garvey is viewed as one of the prophets, as is the late Bob Marley. Followers of the Rastafarian faith have established themselves across language barriers, forming small but visible communities throughout the region, for example in Havana and San Juan. The iconography of Rastafarianism extends much further than the specific focus of its adherents, being widely reproduced for popular consumption in art galleries and boutiques. Formed initially among black intellectuals in Jamaica, the faith spanned countries and classes, and provides a persuasive example of the way in which race and ethnicity can crystallise social movements and impact on popular identities well beyond the locality.

On the move: changing identities

Migration, forced and voluntary, has been a fundamental feature in the development of Middle American societies. From around the tenth century, indigenous trade routes and settlements began to be established that connected points along the mainland isthmus and out to the islands of the Caribbean. Later arrivals linked Europe and the Americas, Africa and Asia as colonisation carved new channels of uneven exchange, trading people, products and capital. Slavery was finally abolished in the region as a whole by the end of the nineteenth century. Prolonged attempts were subsequently made to address the labour shortage through indentureship. Although contracted labour arrived from Europe and Africa, the majority came from Asia: approximately 135,000 Chinese, over half a million Indians and 33,000 indentured workers from Indonesia (Mintz 1974: 313). These indentured labourers settled largely in Trinidad, Suriname and Guyana.

Until the early seventeenth century, around a half of all slaves who arrived from Africa came to Mexico (Purcell 1993). These slaves were considered to be wholly inferior to the indigenous populace, ranked at the absolute bottom of the social hierarchy and often described as evil. The Afro-Mexican presence today is significantly sidelined, the ambiguities of racial definition making numerical estimates difficult or inappropriate. Contemporary figures would suggest that Afro-Mexicans account for 0.5 per cent of the population, although the musical influences of *La Morena*, *La Negra* and *El Maracumbre*, and dances such as the *jarabe*, *gusto* and *zapateo* make an evident connection with African heritage (Muhammad 1996).

Slaves were traded and brought to the mainland Atlantic coast plantations, but numbers were still limited in comparison to the Caribbean plantation economies. Many workers remained and established some of today's African Caribbean communities in Bluefields, Pearl Lagoon Corn Island and Puerto Cabezas on the Niacaraguan coast. African Caribbeans or Creoles further migrated in significant numbers to Central America at the end of nineteenth century to work on the railroad and banana plantations in Costa Rica, Honduras and Panama in particular (Knight 1990; McIlwaine 1997). Employment opportunities on banana plantations and in the mahogany industry at the start of the twentieth century brought additional Caribbean migrants to the coastal zones of the mainland.

The overt violence that scarred Central America during the 1970s and 1980s and displaced hundreds of thousands of people has subsided, reducing a significant source of migration on the mainland. Refugees from civil wars, revolution and state repression had caused thousands of people to flee their homelands. By the mid-1980s, 250,000 Central Americans were seeking refuge in Mexico alone (Fernández-Kelly and Portes 1992: 263). Guerrilla activity arising from the inequalities of wealth, landholdings and limited political representation fuelled conflicts in El Salvador, Nicaragua and Guatemala. By 1992, 8,000 people had been killed in El Salvador as a result of a decade of government-sponsored violence. Over half a million fled the country, while a similar number, approximately 10 per cent of the population, were displaced (Sollis 1992).

Recent waves of migration continue to shape the social and economic profiles of countries in Middle America (Chamberlain 1998). The connectivity of transnational communities expands daily as business and family contacts traverse national boundaries and provide opportunities for employment and interaction. The proximity of the US means that most countries have up to 10 per cent of their population living or working in North America, or at least planning migratory routes Stateside. The US government's *bracero* programme facilitated the emigration of 4 million Mexicans between 1954 and 1964, when legal work contracts were issued in order to address the labour shortages in the southwest of the US. Migrants from the region have made a marked impact at all levels of North American society, from contributing to the economy, arts and politics, to establishing longstanding social networks and neighbourhood connections that span the continent.

Concluding comments

Although regional typologies or general commentaries on ethnicities and nationalism in Middle America may be analytically blunt or fall short of highlighting the subtleties of individual contexts, they provide an entrée to understanding broad regional variations. This chapter aimed to provide an introduction to major themes and to encourage the reader towards more detailed research, which explores the specificities of locality, personal experiences and community histories.

Regional groupings, such as Middle America or the Caribbean, themselves hide the ambiguities of generic labelling. In the case of the Caribbean, inter-governmental organisations such as CARICOM and the Association of Caribbean States seek to foster links between countries, but fail to excite a pan-Caribbean identity beyond the market place. The collapse of the short-lived British West Indies Federation in 1962 illustrates the historical salience of national sentiment, which in many respects negates cohesive notions of the Caribbean.

Decolonisation, independence and the consequent challenge to former power structures occurred across Middle America at varying timescales and with differing results. The dissolution or eruption of new social movements or national identities continues not within the imprint of faltering colonial footsteps, but on the wider expanse of global political, economic and organisational inter-relations. The fusion and shifting of ethnicities in Middle America mirrors the transformations occurring on the larger global stage, the coming together of identities as people, capital and culture migrate across boundaries and through territories. This process is not new, but exists as a perpetuation of the social and ethnic interactions which will continue to shape societies throughout the region.

Further reading

Beckles, H.M.D. and **Shepherd, V.A.** (eds) (1993) *Caribbean Freedom: Economy and Society from Emancipation to the Present*. James Currey, London.

A comprehensive collection of essays covering the formation of contemporary Caribbean societies.

Chomsky, A. and **Lauria-Santiago, A.** (eds) (1998) *Identity and Struggle at the Margins of the Nation-State: the Laboring Peoples of Central America and the Hispanic Caribbean*. Duke University Press, Durham.
This text explains the evolution of national identities and popular conceptions of ethnicity in Middle American societies during the nineteenth and twentieth centuries.

Gutiérrez, N. (1998) What indians say about *mestizo/as*: a critical view of a cultural archetype of Mexican nationalism, *Bulletin of Latin American Research* **17**(3): 285–301.
This article gives voice to indigenous perceptions of the Mexican nation and their exclusion from state-derived ethnic identities.

Kearney, M. and **Stavenhagen, R.** (1996) Special issue: ethnicity and class in Latin America, *Latin American Perspectives* **23**.
A wide-ranging collection of articles with case studies from the Caribbean and Central America.

Oostindie, G. (ed.) (1996) *Ethnicity in the Caribbean*. Macmillan, London.
Ten essays analysing the workings of race and ethnicity and their significance for social structure and national identities across the Caribbean.

References

Anthias, F. (1990) Race and class revisited: conceptualising race and racism, *The Sociological Review* **38**(1): 19–42.

Augelli, J.P. (1989) Cultural characteristics and diversity of Middle America, in West, R.C., Augelli, J.P., Boswell, T.D., Crowley, W.K., Elbow, G.S. and Griffin, E.C. (eds) *Middle America: Its Land and Peoples*. Prentice Hall, Englewood Cliffs: 1–22.

Barry, T. (1992) *Mexico: a Country Guide*. Inter-Hemispheric Education Resource Center, Albuquerque.

Bastide, R. (1961) Dusky Venus, Black Apollo, *Race* **3**: 10–18.

Beckles, H. McD. (1999) Whose game is it anyway? West Indies cricket and post-colonial cultural globalism, in Skelton, T. and Allen, T. (eds) *Culture and Global Change*. Routledge, London: 251–259.

Benítez-Rojo, A. (1992) *The Repeating Island: the Caribbean and the Post-Modern Perspective*. Duke University Press, London.

Cashmore, E. (ed.) (1995) *Dictionary of Race and Ethnic Relations*. Routledge, London.

Chamberlain, M. (1998) *Caribbean Migration: Globalised Identities*. Routledge, London.

Chant, S. (1996) *Gender, Urbanisation and Housing: Issues and Challenges for the 1990s, Environmental and Spatial Analysis (London School of Economics).* Number 32.

Clarke, C.G. (1991) Introduction: Caribbean decolonization – new states and old societies, in Clarke, C. (ed.) *Society and Politics in the Caribbean.* St Anthony's-Macmillan, Basingstoke: 1–27.

Davis, D.J. (ed.) (1995) *Slavery and Beyond: the African Impact on Latin America and the Caribbean.* SR Books, Wilmington.

Edmondson, B. (ed.) (1999) *Caribbean Romances: the Politics of Regional Representation.* University Press of Virginia, Charlottesville.

Esteva-Fabregat, C. (1995) *Mestizaje in Ibero-America.* University of Arizona Press, Tucson.

Fanon, F. (1986) *Black Skin, White Masks.* Pluto Press, London.

Fernández Kelly, M.P. and **Portes, A.** (1992) Continent on the move: immigrants and refugees in the Americas, in Stepan, A. (ed.) *Americas: New Interpretive Essays.* Oxford University Press, New York: 248–274.

Gall, O. (1998) Los elementos histórico-estructurales del racismo en Chiapas, in Castellanos-Guerrero, A. and Palacios, J.M.S. (eds) *Nación, Racismo e Identidad.* Editora Nuestro Tiempo, Mexico City: 43–190.

Gould, J. (1998) *To Die in This Way: Nicaraguan Indians and the Myth of Mestizaje, 1880–1965.* Duke University Press, London.

Gutiérrez, N. (1998) What indians say about *mestizo/as*: a critical view of a cultural archetype of Mexican nationalism, *Bulletin of Latin American Research* **17**(3): 285–301.

Hale, C.R. (1994) *Resistance and Contradiction: Miskitu Indians and the Nicaraguan State, 1894–1987.* Stanford University Press, Stanford.

Hall, S. (1990) The whites of their eyes, in Alvarado, M. and Thompson, J.O. (eds) *The Media Reader.* British Film Institute, London: 7–23.

Helg, A. (1995) *Our Rightful Share: the Afro-Cuban Struggle for Equality, 1886–1912,* University of North Carolina Press, Chapel Hill.

Hettne, B. (1995) *Development Theory and the Three Worlds.* Longman, Harlow.

Hoetink, H. (1985) 'Race' and colour in the Caribbean, in Mintz, S. and Price, S. (eds) *Caribbean Contours.* The Johns Hopkins University Press, Baltimore: 55–85.

Hondagneu-Sotelo, P. (1992) Overcoming patriarchal constraints: the reconstruction of gender relations among Mexican immigrant women and men, *Gender and Society* **6**(3): 393–415.

Howard, D.J. (1998) *The Dominican Republic: a Guide to the People, Politics and Culture.* Latin American Bureau, London.

Jackson, P. (1994) Black Male: advertising and the cultural politics of masculinity, *Gender, Place and Culture* **1**(1): 49–59.

Jackson, P. and **Penrose, J.** (1993) *Constructions of Race, Place and Nation.* University College London Press, London.

Knight, A. (1990) Racism, Revolution, and *Indigenismo*: Mexico, 1910–1940, in Graham, R. (ed.) *The Idea of Race in Latin America, 1870–1940.* University of Texas Press, Austin: 71–113.

Knight, F.W. (1990) *The Caribbean: the Genesis of a Fragmented Nationalism* (2nd edn.). Oxford University Press, New York.

Kroshus Medina, L. (1997) Defining difference, forging unity: the co-construction of race, ethnicity and nation in Belize, *Ethnic and Racial Studies,* **20**(4): 757–780.

Lancaster, R.N. (1992) *Life is Hard: Machismo, Danger, and the Intimacy of Power in Nicaragua.* University of California Press, Berkeley.

Lowenthal, D. (1972) *West Indian Societies.* Oxford University Press, Oxford.

McIlwaine, C. (1995) Gender, race and ethnicity: concepts, realities and policy implications, *Third World Planning Review* **17**(2): 237–244.

—— (1997) Vulnerable or poor? A study of ethnic and gender disadvantage among Afro-Caribbeans in Limón, Costa Rica, *The European Journal of Development Research* **9**(2): 35–61.

Mintz, S.W. (1974) *Caribbean Transformations.* Aldine, Chicago.

Moberg, M. (1997) *Myths of Ethnicity and Nation: Immigration, Work and Identity in the Belize Banana Industry.* University of Tennessee Press, Knoxville.

Morris, S.D. (1999) Reforming the nation: Mexican nationalism in context, *Journal of Latin American Studies* **31**(2): 363–398.

Muhammad, J.S. (1996) Mexico, in Minority Rights Group (ed.) *Afro-Central Americans: Rediscovering the African heritage.* Minority Rights Group, London: 8–14.

Nicholls, D. (1996) *From Dessalines to Duvalier: Race, Colour and National Dependence in Haiti.* Macmillan, Basingstoke.

Oostindie, G. (1996) Introduction: ethnicity, as ever?, in Oostindie, G. (ed.) *Ethnicity in the Caribbean: Essays in Honor of Harry Hoetink.* Macmillan, Basingstoke: 1–21.

O'Kane, T. (1999) *Guatemala: a Guide to the People, Politics and Culture.* Latin American Bureau, London.

Purcell, T. (1993) *Banana Fallout: Class, Color, and Culture among West Indians in Costa Rica.* Center for Afro-American Studies, University of California, Los Angeles.

Radcliffe, S.A. (1999) Embodying national identities: *mestizo* men and white women in Ecuadorian racial-national imaginaries, *Transactions of the Institute of British Geographers* **24**: 213–225.

Recovery of Historical Memory Project (REHMI) (1999) *Guatemala: Never again*. CIIR and Latin American Bureau, London.

Russell, P.L. (1995) *The Chiapas Rebellion*. Mexico Resource Center, Austin, TX.

Sollis, P. (1992) Displaced persons and human rights: the crisis in El Salvador, *Bulletin of Latin American Research* **11**(1): 46–67.

Thornberry, P. (1991) *Minorities and Human Rights Law*. Minority Rights Group, London.

Vasconcelos, J. (1920) *La Raza Cósmica*. Agencia Mundial de Librería, Mexico City.

Wagley, C. (1968) *The Latin American Tradition*. Columbia University Press, New York.

Walby, S. (1990) *Theorizing Patriarchy*. Blackwell, Oxford.

Warren, K.B. (1998) *Indigenous Movements and Their Critics*. Princeton University Press, Princeton, NJ.

Wearne, P. (1994) *The Maya of Guatemala*. Minority Rights Group, London.

—— (1996) *Return of the Indian: Conquest and Survival in the Americas*. Latin American Bureau, London.

Williams, E. (1944) *Capitalism and Slavery*. University of North Carolina, Chapel Hill.

Worsley, P. (2000) Culture and development theory, in Skelton, T. and Allen, T. (eds) *Culture and Global Change*. Routledge, London: 30–41.

CHAPTER 4

Perspectives on poverty, vulnerability and exclusion

Cathy McIlwaine

Introduction

> 'For me, being poor is having to wear trousers that are too big for me.'
> (José, 8 years old, Guatemala City)

> 'Poverty makes my children get sick and they get worse because we're too poor to buy medicines.'
> (Antonia, 30 years old, Esquipulas, Guatemala)

> 'It's poverty that makes me drink until I fall over, and drinking until I fall over makes me poor.'
> (Eduardo, 35 years old, San Marcos, Guatemala)[1]

As these quotations from the poor in Guatemala illustrate, poverty means different things to different people. This chapter explores the dynamics of poverty and related concepts, considering the multidimensionality of poverty, both in theory and in practice. Specifically, the chapter argues for the need to examine poverty from a perspective that moves beyond income and incorporates more subjective elements of deprivation reflected in the use of the terms 'vulnerability' and 'social exclusion'. This is especially pertinent in the case of Middle American countries that have suffered widespread poverty and exclusion both historically and today, and often to a greater extent than elsewhere in Latin America. Poverty in the region

---

[1] These quotations are taken from a research project in Guatemala (and Colombia) conducted in 1999 using Participatory Urban Appraisal (PUA) techniques under the auspices of the Urban Peace Program directed by Caroline Moser and funded by the Swedish International Development Co-operation Agency (SIDA). Although the research was primarily focused on urban violence, its inductive nature meant that poverty issues also emerged (see Moser and McIlwaine 1999, 2000, 2001).

has also been particularly adversely affected by the recession of the 1980s associated with structural adjustment programmes (SAPs). These involve the reorientation of economies along free-market lines under the direction of the World Bank and International Monetary Fund (IMF). Although the exact relationship between SAPs and poverty is contentious, adjustment involves severe cutbacks in social spending that have deleterious effects on the well-being of populations.

Following a discussion of poverty concepts with reference to Middle America, the chapter examines the dynamics of poverty, outlining some key patterns and trends in the region. In particular, it highlights rural–urban differences, and some basic characteristics of those identified as poor in the region. It then considers the reasons for the changes in the incidence of poverty, vulnerability and exclusion in the region with specific reference to economic recession and SAPs.

Conceptual perspectives on poverty, vulnerability and exclusion

What is poverty?

It is now widely accepted that the concept of poverty is contested, often depending on the disciplinary viewpoint adopted, as well as on who is responsible for its definition – the poverty 'expert' or the poor themselves. While debates on poverty are certainly not new, recent research has attracted much greater and more interdisciplinary attention (Pinker 1999). Indeed, over the latter half of the last century, poverty gradually became a central concern within development debates (Thomas 2000). By the year 2000, a series of poverty reduction targets and pledges were in place within the development community. Multilateral and bilateral development organisations and agencies, such as the World Bank, United Nations Development Programme (UNDP), and the UK Department for International Development have all adopted poverty reduction, or even elimination, as a key element of their respective strategies. Indeed, in a follow-up to its 1990 version, the World Bank's 2000/2001 *World Development Report* focuses on 'Attacking Poverty' (World Bank 2000). Following the UN Social Summit in 1996, there is now consensus among these agencies, as well as national governments, to commit to reduce the proportion of people living in extreme poverty by one half by 2015. An important aspect of recent changes has been the Highly Indebted Poor Country (HIPC) debt relief initiative launched by the World Bank and IMF in 1996 (HIPC I) and reviewed and modified in 1999 (HIPC II). This initiative has made debt relief in 41 of the poorest countries in the world conditional on the creation of a national poverty reduction framework. This framework addresses the nature of poverty in each country as well as policies and budgets to implement them, and is outlined in a Poverty Reduction Strategy Paper (PRSP) prepared by the country in collaboration with the World Bank, IMF and civil society (Healey *et al.* 2000; World Bank 2000: 201). In Middle America, Guyana, Honduras and Nicaragua have all been involved in this initiative, representing three out of a total four Latin American countries included (UNICEF/Oxfam 1999).

At the risk of oversimplifying an extremely complex issue, poverty thus refers to a broad spectrum of deprivation and disadvantage (Baulch 1996). Pinning down what poverty specifically means, however, is more difficult. Rahnema (1995: 8) goes as far as to say that '[p]overty is also a myth, a construct and the invention of a particular civilization . . . There may be as many poor and as many perceptions of poverty as there are human beings.' Various different types of poverty have been identified, with some specifically relevant to Middle America. These include 'structural poverty' that is permanent and less open to vagaries of the economy, and 'new poverty' arising from the debt crisis and SAPs (see below) (Pérez Sáinz 1998). Ottone (1997) adds 'poverty of the marginalized' found among those with no access to material goods over time, with poverty being passed from one generation to the next, and 'regional poverty' denoting spatially concentrated poverty as a result of regionally unequal development. Besides highlighting the diversity of different types of poverty, these also illustrate how poverty is dynamic and multi-dimensional (Chambers 1995).

However, the complexities extend beyond these differences, both in Middle America and further afield. Maxwell (1999) identifies a series of so-called 'fault lines' in the debates on poverty. These highlight variations in individual or household measures, monetary or non-monetary components of poverty, and objective or subjective perceptions of poverty. Within these 'fault lines' there are two broadly different ways of conceptualising poverty. The first focuses on measuring income and consumption, drawing heavily on the use of poverty lines measured by statistics. The second emphasises subjective interpretations of poverty relying on participatory methods focusing on what people regard as poverty (see also Watkins 1995).

Does poverty mean money and spending? Income and consumption approaches

Since the middle of the last century, efforts have been made to measure poverty according to income and consumption. These quantitative definitions of poverty tend to concentrate on the measurement of poverty lines. A poverty line generally refers to a threshold in terms of wealth below which people can be considered as poor. Although different countries use different approaches, poverty lines can be calculated either according to estimates of income in terms of command over resources referring to what is earned, or as a level of consumption defined as what needs to be consumed or spent in order to survive. Therefore, using household survey data, poverty lines estimate a level of income or expenditure necessary for buying sufficient food to satisfy the average nutritional needs of everyone within a household. The cost of this food is the basic cost of subsistence, sometimes called the 'basic basket'. When this is added to an allowance for basic clothing, fuel and rent, a figure is estimated below which a household is said to be living in poverty. When people fail to achieve the minimal acceptable amount of income or consumption, then they are seen to be in poverty (Webster 1990; Wratten 1995).

As the most commonly used method of defining welfare and standards of living, poverty lines reflect absolute measures of poverty. For instance, the World Bank developed two international poverty lines in an attempt to provide cross-country comparisons in its 1990 *World Development Report*. It used two income poverty lines; the first identified those with an income per capita of below US$370 per year (using 1985 adjusted prices), who are deemed poor, while those with less than US$275 are extremely poor (World Bank 1990). The 2000/2001 *World Development Report* updated these poverty lines using the standard as US$1.08 a day and US$2 per day based on 1993 adjusted international prices (World Bank 2000) (see also Table 4.1, page 91).[2] However, there are also technical measures that attempt to include relative estimates of poverty in a statistical sense. The most common are the headcount index and the poverty gap. The former measures the proportion of people in poverty estimated by the percentage living below a set poverty line; the latter measures the average distance below the poverty line expressed as a percentage of the poverty line; this identifies the depth or severity of poverty and effectively measures the amount of resources needed to bring those below the poverty line to a reasonable living standard (Baden and Milward 1995). In an example from Middle America, the headcount index in Guyana (based on US$1 per day) was 43.2 per cent in 1993. In turn, the poverty gap stood at 16.2 per cent (Gafar 1998: 608).

Although poverty lines are useful for making cross-country comparisons, especially for policymakers, they have been widely criticised. Among the most important criticisms is that income or consumption patterns only allow identification of who lacks resources, but not the capacity to achieve access to resources. This is sometimes referred to as analysis of both the 'ends' and the 'means' of poverty, with 'means' being especially influenced by gender, ethnicity and class (Kabeer 1996). Also significant from a gender perspective is that poverty lines are usually measured at the household level, yet individual members of households do not necessarily have the same access to resources within households, especially women and children (see below). In turn, poverty lines do not account for household size, therefore failing to consider economies of scale from living in larger households, or for income not included in national accounts, such as informal sector activity. A final set of criticisms is that cultures and customs vary so widely across the South that it is impossible to make accurate predictions or assumptions about peoples' needs and the distribution of income and expenditures within households. These also highlight how poverty depends on social norms and expectations (Wratten 1995; UNCHS 1996).

In the light of these criticisms, efforts have been made to extend indicators beyond poverty lines to include other variables such as education or health indicators. The most well known is the UNDP's Human Development Index (HDI) that measures income, literacy and life expectancy combined in a single index (UNDP 1990). Variations on the HDI have been developed over the last decade by the UNDP, such as the Capability Poverty Measure

---

[2] These adjusted figures are international prices calculated using Purchasing Power Parity (PPP) which converts local prices and costs within countries into internationally comparable ones in a given year (in 1985 and 1993 in the case of the World Bank) (World Bank 2000: 320).

that estimates the percentage of people who lack basic or minimally essential human capabilities in relation to health, reproduction and education (UNDP 1996). A more recent variation is the Human Poverty Index. This is another composite deprivation index that brings together four dimensions – a long and healthy life, knowledge, economic provisioning and social inclusion (UNDP 1999: 130). Caribbean countries, in particular, fare very well according to this measure; of a total of 92, Barbados and Trinidad and Tobago were ranked in first and second place in 1999 (see Table 4.2 on page 92). While overall, these are certainly improvements in poverty line measurements, they are still subject to similar criticisms in terms of their technocratic approach and neglect of the views of the poor themselves. However, they provide useful comparisons at a global and regional scale.

Beyond income: subjective and participatory interpretations of poverty

Emerging from criticisms of static definitions of poverty based on technical measurements have been a series of qualitative approaches to conceptualising poverty. These attempt to incorporate more subjective elements of the experiences of poverty, and/or the perceptions of the poor themselves. Broadly speaking, they rest on broader and more dynamic interpretations of poverty that include 'social inferiority, isolation, physical weakness, vulnerability, seasonal deprivation, powerlessness and humiliation' (Chambers 1995: 3).

These aspects of poverty have frequently arisen from participatory appraisal methodologies that are conducted with poor people themselves, asking them how they themselves define poverty (Chambers 1997) (see below on examples from Guatemala). Often incorporated into participatory poverty assessments (PPAs), championed by the World Bank in particular, this approach has continually revealed how the poor themselves often define poverty in terms of self-respect, autonomy, and access to assets such as land rather than income. A World Bank PPA undertaken in Costa Rica in 1991, for example, illustrated the links between home ownership and status in society, rather than income and status (Robb 1999: 97). Another in urban Jamaica, exploring poverty and violence, highlighted how 'poor people' lacked good housing and employment and had financial difficulties and lots of children. Those thought to be 'very, very poor people' were a blind lady, a beggar, the elderly and those who live in a shack (Moser and Holland 1997: 6). One of the largest projects undertaken to date using PPAs is the 'Voices of the Poor' study conducted by the World Bank for the 2000/2001 *World Development Report*. Carried out in 60 countries, this highlights the importance of participatory conceptualisations in understanding poverty (World Bank 2000: 16).

Although these participatory approaches have been criticised because they are difficult to measure (Baulch 1996), and because they are sometimes thought to neglect gender issues, they have flourished in the light of a series of recent concepts that distinguish income poverty from other dimensions of disadvantage. Underlying much of this reconstitution of poverty is the issue of entitlements and capabilities, which draws upon research by Amartya Sen in particular, on food security and famines in rural areas (Sen 1981). The importance of this approach is recognition of how the poor obtain as well as command resources through the use of entitlements in order to withstand short-term shocks and longer-term trends. These entitlements may refer to income from wage labour, income from the sale of assets, resources from own production such as food, social security claims, assets and so on. Those able to access entitlements during times of crisis are often more able to guard against deprivation (Amis and Rakodi 1994).

The notion of entitlements also underlies the concept of vulnerability. As with the former, the term allows a dynamic analysis of disadvantage that extends beyond income poverty (Moser 1998). The nature of vulnerability lies in how the poor develop coping strategies in order to withstand shocks through diversifying and mobilising their asset base. Their asset base may include labour, human capital (such as health and education), productive assets (such as land and housing), household relations (focusing on income pooling and consumption sharing), and social capital (see Box 4.1)

Box 4.1   What is social capital?

Social capital is defined as the rules, reciprocity and trust embedded in social relations, social structures, and social institutions. Social capital is an asset that can be drawn upon in times of need, for example, turning to networks of friends or neighbours to borrow money, to assist with childcare or to help in job placement. People are therefore able to secure benefits from their membership in social networks or social institutions. In this way, it can help guard against falling into poverty or dealing with crisis.

There are many different types of social capital, but one of the most important distinctions is between *cognitive* and *structural social capital*. Structural social capital incorporates interpersonal relationships within organisations or networks, while cognitive social capital refers to informal networks of reciprocity, as well as attitudes and beliefs such as trust and fear.

Another useful distinction is between *horizontal* and *vertical social capital*. Horizontal refers to those relationships that occur among people and institutions on a similar level, such as community networks among people of roughly similar status. Vertical social capital is hierarchical in structure, and involves relations among people or organisations of unequal status, such as between a community group and a large non-governmental organisation (NGO).

A final distinction is between *productive* and *perverse social capital*. Productive social capital generates favourable relationships with positive outcomes – ensuring that a community is united in the face of a natural disaster. Perverse social capital is associated with negative outcomes with benefits only for those within perverse institutions. A perverse institution would be a gang that is often internally cohesive but is involved in violence and crime to the detriment of a community.

Sources: Fine (1999); Harriss and De Renzio (1997); Krishna and Uphoff (1998); Moser and McIlwaine (2001); Portes (1998); Rubio (1997).

(Moser 1998; see also Bebbington 1999). People are vulnerable when they have few assets on which to depend to prevent hardship; moreover, they may be relatively well off in income terms, yet vulnerable in other ways. These may include exposure to environmental risks, such as flooding (see Pelling 1997 on vulnerability and floods in Guyana), earthquakes, crime, violence, or social exclusion (UNCHS 1996: 108–9). Gender and ethnic dimensions of vulnerability are especially important here, in that women and ethnic minority groups, for example, may be vulnerable yet may not be poor in the income sense. In Georgetown, Guyana, for instance, the Afro-Guyanese population are vulnerable due to discrimination, especially in terms of their mobility through the labour market, yet they are not necessarily the poorest in income terms (Trotz 1996; see also Box 4.2 on Costa Rica).

Social exclusion is another notion that has been increasingly used to express the subjective dimensions of poverty that are not captured by poverty lines. The term was originally coined in France in the 1970s to conceptualise disadvantage linked with those who were not covered by state social security, and especially by employment generation schemes. There are four key characteristics of social exclusion that separate it from income poverty measures. The first is that social exclusion is multidimensional in that it extends beyond the economic sphere of income poverty to include social and political arenas as well as power relations, social agency and identity. Second, it is dynamic, referring to the processes or mechanisms by which people are excluded. Third, it acknowledges the central role of institutions and actors within processes of exclusion. Finally, social exclusion is the opposite of social inclusion or integration, reinforcing the importance of being part of society rather than just lacking access to a range of resources (de Haan 1998: 12–13). Simply then, it refers to the process through which individuals or groups are wholly or partially excluded from full participation in the society in which they live. This exclusion may be economic, social or political (de Haan 1998; de Haan and Maxwell 1998).

Recently, the concept of social exclusion has become widespread, not only in Europe, but also in countries in the South. Indeed, it is now widely incorporated into poverty debates in developing countries, most prominently by the International Institute for Labour Studies at the International Labour Office (ILO) (IILS/UNDP 1996). It has been most commonly used in relation to policy work rather than pure research, with many international organisations using social exclusion as a key policy concern; it is frequently seen to be more comprehensive than a strict focus on poverty. Therefore, although it has been criticised as a Western concept applied to non-European countries, it has, none the less, been seen as a useful way of examining the multidimensionality of deprivation. Furthermore, it also highlights the ways in which poverty changes according to dominant social norms in a particular country.

A series of dimensions were identified in a Central American study on social exclusion under the auspices of the UNDP in Costa Rica, El Salvador and Guatemala. These were then transformed into a range of direct and indirect indicators that were subsequently calculated and used to measure social exclusion. The direct indicators included income poverty, child

Box 4.2   Poverty and vulnerability in Costa Rica: a case study of Afro-Caribbeans in Limón

The city of Limón developed at the turn of the century as the apex of an enclave economy largely dependent on banana cultivation. It was established and grew in response first to the demands of railway construction from the centre of the country to the Atlantic coast as a transport route for coffee, and second to the needs of the United Fruit Company, both occurring before the turn of the twentieth century. The province and city are characterised by large proportions of Afro-Caribbeans, who migrated mainly from Jamaica to work on the railway or banana plantations. The Afro-Caribbean population has had a chequered history in the region, being a privileged workforce under United Fruit, yet becoming a discriminated against ethnic minority in the eyes of the Costa Rican state.

The concept of vulnerability is more useful in this case than income poverty for assessing their position in the contemporary labour market. Because of their important role in the economy of the city and province, Afro-Caribbeans tend to have lower levels of poverty than white/*mestizo* groups; for example, in a household survey conducted in 1990, of 80 low-income Afro-Caribbean and 170 white/*mestizo* households, Afro-Caribbeans – especially women – had higher per capita and total household incomes than their *mestizo* counterparts, as well as higher average earnings. However, although Afro-Caribbeans were more privileged than white/*mestizos* in a narrow sense of material advantage, they were discriminated against in other ways that made them vulnerable.

The vulnerability of Afro-Caribbeans was most obvious in relation to ethnic occupational divisions and employment mobility. Most important is the inability of this group to transcend the traditional occupational segmentation and segregation that have been entrenched since the days of United Fruit. For example, both male and female Afro-Caribbeans have been employed in the same sectors of the labour market since the days of United Fruit – in port handling for men and public sector nursing and teaching for women. Both sexes have remained virtually excluded from key decision-making positions in the upper levels of the employment structure that is dominated by white/*mestizo* men. The only way that they can break free of these occupational divisions is to migrate to the US – a common strategy among this population.

Therefore, the situation of Afro-Caribbeans is more accurately assessed by examining their vulnerability. While this group have historically been able to guard against economic vulnerability in terms of income, they have experienced labour vulnerability through occupational entrenchment in ways that have been both racialised and gendered. Afro-Caribbean vulnerability is also related to powerlessness and social exclusion within the national polity.

Source: McIlwaine (1997).

Plate 4.1   Vulnerability in Guatemala City: homeless urban poor living in a
primary school immediately after hurricane 'Mitch'
Photo: Cathy McIlwaine

malnutrition, illiteracy and territorial isolation (among others). The indirect
indicators referred to economic precariousness and discrimination including
ethnic, linguistic and political-marginality of women, lack of access to
health and sanitation, cultural discrimination, and social abandonment.
Guatemala emerged as suffering the most social exclusion, followed by El
Salvador and then Costa Rica. However, perhaps the most significant
aspect of the study was the emphasis on the dynamism and multifaceted
aspects of social exclusion as a concept, allowing for a more rounded
interpretation of more intangible characteristics of countries, such as lack
of social cohesion and human dignity (Menjívar and Feliciani 1995).

The nature and extent of poverty in Middle America

How widespread is poverty?

Given the difficulties with presenting cross-country comparisons using
participatory or subjective approaches to poverty, this section focuses
mainly on national level quantitative measurements, emphasising income
poverty. Also it should be highlighted that the ways in which poverty is

Table 4.1 National and international poverty lines in selected Middle American countries

| | *Proportion of population living below the poverty line* | | | | | | |
| | *National poverty lines* | | | | *International poverty lines* | | |
Country	*Survey year*	*Rural*	*Urban*	*National*	*Survey year*	*Percentage population below $1 per day*	*Percentage population below $2 per day*
Costa Rica					1995	9.6	26.3
Dominican							
Republic	1989	27.4	23.3	24.5	1996	3.2	16.0
El Salvador	1992	55.7	43.1	48.3	1996	25.3	51.9
Guatemala	1989	71.9	33.7	57.9	1989	39.8	64.3
Honduras	1992	46.0	56.0	50.0	1996	40.5	68.8
Jamaica	1992			34.2	1996	3.2	25.2
Mexico	1988			10.1	1995	17.9	42.5
Nicaragua	1993	76.1	31.9	50.3			
Panama	1997	64.9	15.3	37.3	1997	10.3	25.1
Trinidad and							
Tobago	1992		21.0				

Source: World Bank (2000: 280–281).

measured vary from country to country. Within Middle America, some countries – such as Costa Rica, Panama, El Salvador and Guatemala – rely on poverty lines based on income and define the poor as those unable to buy a 'basic basket' of foodstuffs. In contrast, other countries such as Honduras and Nicaragua rely on a 'basic needs' approach that denotes poverty as equated with non-satisfaction of necessities that include basic goods and services as well as food (Menjívar and Trejos, cited in Chant 1997a: 67).

As measured by a mixture of different poverty lines, poverty in Middle America as a whole is higher than in South America, especially compared with the Southern Cone countries of Argentina and Uruguay (Gilbert 1997: 322). However, within Middle America, Central American nations tend to be poorer than those in the Caribbean. In many Central American countries, more than 50 per cent of households live in poverty. For instance, in Nicaragua, which is the poorest country in Latin America (Oxfam 1998), in 1993, 43.8 per cent lived below US$1 per day, with 74.5 per cent subsisting on less than US$2. In turn, figures for some Caribbean countries are better; in Jamaica in 1992, 4.3 per cent lived below the US$1 poverty line, with 24.9 per cent living below US$2 (see Table 4.1). Overall, nearly 38 per cent of the total population in the Caribbean have been classified as poor; if Haiti – the poorest country in the Caribbean – is excluded, this figure drops to 25 per cent (Gafar 1998: 608). Having said this, poverty levels are still of major concern in the region; even if the proportion of Jamaica's

Table 4.2 Human Poverty Index (of developing countries) in selected Middle American countries

Country	Country Rank	Human Poverty Index (1997)
Barbados	1	2.6
Costa Rica	4	4.1
Cuba	5	4.7
Dominican Republic	26	17.7
El Salvador	35	20.6
Guatemala	50	28.3
Guyana	11	10.2
Honduras	41	25.0
Jamaica	17	13.6
Mexico	13	10.6
Nicaragua	48	28.1
Panama	8	9.0
Trinidad and Tobago	2	3.5

Source: UNDP (1999: 146–147).
Note
The country rank refers to the developing world only

population living in poverty is considerably less than in Guatemala, there remains one-quarter of all people living on less than US$2 per day.

Furthermore, it should also be noted that these estimates vary widely, and caution should therefore be exercised. While all these figures depend on World Bank statistics, other sources cite different poverty lines for the same countries; for example, for Costa Rica the World Bank suggests that in 1989, 43.8 per cent of households were poor, while Altimir (1994: 11–12) using different statistics says that 24 per cent were poor.

However, when assessed according to the UNDP's Human Poverty Index, the picture is slightly better. Among the 92 countries for which the Human Poverty Index is calculated, those in Middle America fare relatively well; indeed, Barbados has the lowest level of human poverty of all nations included, with Costa Rica, Panama, Cuba and Trinidad and Tobago all with low indices (UNDP 1999: 31). This would therefore suggest that social service facilities, such as education and healthcare, are relatively well developed in these nations. None the less, figures for the Human Poverty Index also highlight the huge diversity within Middle America. As with the patterns for income poverty, Nicaragua and Guatemala are ranked in 48th and 50th position respectively (see Table 4.2).

Who are the poor?

The diversity in terms of poverty is also reflected within nations. A major axis of differentiation is between rural and urban areas. As elsewhere in the world, rural poverty in Middle America is generally more widespread and severe than in urban areas. For Latin America as a whole, for instance, Gilbert (1997: 323) suggests that 39 per cent of city dwellers live in poverty compared with 61 per cent of people living in the countryside. These differentials are more severe within some countries than in others. For example, in Nicaragua in 1993, 76.1 per cent of rural households lived below the poverty line compared with 31.9 per cent of urban families. In contrast, in the Dominican Republic, the difference was much less pronounced; 27.4 per cent of the population in rural areas and 23.3 per cent in urban areas lived below the poverty line (see Table 4.1). Having said this, subjective interpretations also illustrate that poverty experienced in urban areas is often just as severe as in rural areas. Indeed, as Middle American countries become more urbanised, urban poverty is becoming ever more serious (see below).

These poverty levels are further compounded by gaps in access to services between rural and urban areas. In Mexico, where 23 per cent of the urban population and 43 per cent of the rural population are poor, 34 per cent of the rural population still lack access to safe water whereas only 6 per cent of the urban population face such difficulties. In Honduras, the differential is even greater; even though poverty levels are more or less uniformly high in rural and urban areas (74 per cent in urban areas and 80 per cent in rural), only 51 per cent of the rural population has access to safe water compared with 89 per cent of urban dwellers (UNDP 1995: 166–167). However, having said this, in Middle America there are more urban than rural poor in terms of absolute numbers (Gafar 1998). This pattern has become more marked since the 1980s due to economic recession (see below).

While these figures reflect mainly income poverty, it is interesting to consider a participatory example of what urban poverty means for the poor themselves. Based on Participatory Urban Appraisal (PUA) techniques (Moser and McIlwaine 1999), Figure 4.1 highlights how a group of six men

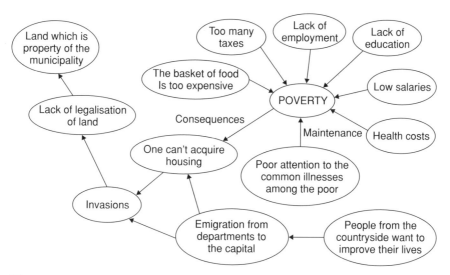

Figure 4.1 Causal impact diagram of poverty in Guatemala City, drawn by a group of six adult men in a local community hall

in an urban poor community in Guatemala City perceived poverty. Although they make one reference to the 'basket of food' referring to average household food costs, as well as to 'low salaries', they also associate poverty with lack of education, health costs and difficulties in accessing housing, especially in terms of legal land tenure. They also identify the links between rural and urban poverty, highlighting how people migrate from the countryside to escape poverty, yet also face problems in the city. One of the men in this group said that rural poverty was actually more severe than urban in his view, especially as it was difficult to access land (Guatemala has one of the region's most unequal distributions of land).

Another important axis is gender. Women are generally perceived to be poorer than men, both in Middle America and elsewhere. This is mainly because women lack the entitlements enjoyed by men and are more constrained in transforming their labour into income in terms of undertaking paid work (Kabeer 1996, 1997). However, it is extremely difficult to assess gender differentials in poverty due to a severe lack of gender-disaggregated data. Because poverty lines are constructed using household survey data, gender differences in experiences of poverty within households are ignored. Due to these data problems, most estimations of the extent of poverty among women tends to use female household headship as a proxy for proportions of women living in poverty. This assumption is made on the basis that women-headed households are likely to have fewer wage earners than two-parent units, women tend to be concentrated in inferior and lower-paying jobs, exacerbated by women having to perform more domestic responsibilities (Baden and Milward 1995; Chant 1997a).

However, although this relationship between female-headed households and poverty is relevant in some contexts, it is not always the case (Chant 1997b). Indeed, it has been repeatedly shown in Middle America and elsewhere that this relationship should not be assumed (see Chant, this volume). In a study of poverty in Central America, for example, only Nicaragua and El Salvador had higher levels of female headship among the poor than the national average (Menjívar and Trejos, cited in Chant 1997a: 49). Moreover, a wide range of empirical examples from the region highlight how female-headed households are often better off than male-headed, as was the case in a study on Panama (Fuwa 2000). Similarly, in research from Mexico and Costa Rica, Chant (1997b: 37–38) shows that women-headed households had higher per capita incomes (total household earnings divided by household size) than male-headed units. In the same cases, Chant (1997a, 1997b) also highlights the importance of 'secondary poverty' within male-headed families whereby men withhold earnings from female spouses, often for various leisure pursuits. Indeed, in another study of Mexico, it has been shown that men may withhold up to 50 per cent of their earnings (González de la Rocha, cited in Chant 1996a: 14). Therefore, women-headed units are free from this form of hidden poverty. A final danger of associating women with poverty via female household headship is that it cannot be assumed that these families are only an outcome of poverty; there are other demographic, cultural and institutional factors involved in their formation (Chant 1997a; see also Chant, this volume).

This association of women-headed households with poverty, coupled with women's concentration in low status, informal sector occupations, has led many to argue that there has been a 'feminisation of poverty' throughout the developing world (Baden and Milward 1995). However, caution must be exercised when discussing this notion because, first, it cannot be assumed that because the poor are mostly female then by investing in women, poverty will be reduced; second, it is dangerous to suggest that poverty reduction will eradicate gender inequalities. Instead, it should be recognised that living in poverty is a gendered experience rather than a female-only experience and that women's subordination is not derived only from poverty (Jackson 1996). In Barbados, for example, over 40 per cent of all households are headed by women, yet the nation is relatively well developed in terms of GNP and its Human Poverty Index ranking (Chant and McIlwaine 1998; see also above). Indeed, there are more female-headed households here than in other countries with a much lower GNP and worse poverty indicators.

Overall, however, it is unfortunately still the case that women in Middle America suffer greater material disadvantage than men. In all the countries in the region, women earn less than men. Although women are the mainstay of agricultural trading in St Vincent and the Grenadines, they earn less than men; women agricultural workers are paid a daily wage of US$2.7 compared with US$10.4 for men (Chant and McIlwaine 1998: 23). Likewise, in Barbados women's share of earned income (39.4 per cent) is still lower than their share of the labour force (ibid.: 46). As well as income poverty, women suffer more than men from other aspects of deprivation associated with differential access to education, healthcare and political participation, all rooted in men's preferential access to power in the region. From a quantitative perspective, this has been measured using the Gender-related Development Index (GDI), developed by the UNDP in 1995, which measures the share of earned income between men and women, life expectancy, adult literacy and combined primary, secondary and tertiary educational enrolment. Of the countries for which the index has been calculated, Caribbean countries in particular are ranked quite high (indicating relatively low levels of gender inequalities), with Barbados, Antigua and Barbuda and Trinidad and Tobago in the top 50 ranks globally. The other Middle American countries that fare quite well include Costa Rica, Panama and Mexico, also within the top 50. Those in the region with low indices, and hence high levels of gender inequalities, include Nicaragua (ranked 121st), Guatemala, and Honduras (UNDP 1999: 138–140). It should also be noted that a low rank on GDI does not necessarily mean that women are being more discriminated against than higher ranking states. The low ranking could also be due to a low HDI to start with.

Another major axis when considering the characteristics of the poor in Middle America, is race and ethnicity (see also Howard, this volume). In particular, indigenous Indian populations suffer disproportionately from poverty. In Mexico, the states with the highest proportions of indigenous peoples – Chiapas, Oaxaca and Hidalgo – are also the poorest, while the richest – Nuevo León, Baja California and the Distrito Federal (in Mexico City) – have a small indigenous presence (Gilbert 1997: 324). Similarly, in

Guyana poverty is highest among the Amerindians (indigenous); indeed, 87.5 per cent of this group live below the poverty line, compared with 43 per cent of Afro-Guyanese and 33.7 per cent of Indo-Guyanese (those of Asian or Middle Eastern origin). Furthermore, most Amerindians lived in the rural interior of the country, while the other two groups were more likely to live in urban areas (Gafar 1998: 611–612). In Panama, indigenous populations, also concentrated in rural areas, have the highest concentrations of poverty in the country (Fuwa 2000).

As with gender, ethnic inequalities extend beyond income poverty in Middle America (see also Box 4.2 on page 89). Indigenous peoples in particular have suffered from centuries of social exclusion linked with colonialism and capitalist development in the region. In many countries, these inequalities have underlain civil wars and conflict, especially in countries such as Guatemala and Mexico (Howard, this volume; Moser and McIlwaine 2001). Guatemala probably has one of the worst experiences of active and explicit exclusion of indigenous people, resulting in the widespread massacre of thousands during the worst years of the civil war in the 1980s (REHMI 1999; Howard, this volume). While the signing of peace accords in 1996 also involved guarantees of addressing the severe inequalities that were social and political as well as economic, differentials remain widespread. A UNDP country report on Guatemala found that the HDI was much lower in departments dominated by indigenous peoples – El Quiché, Huehuetenango, San Marcos – in predominantly rural areas. However, the report also discovered that there were differences among indigenous groups; the Q'eqchí were much worse off than the K'iché, Mam and Kakchikel groups (PNUD 1998: 22).

Drawing again from the PUA in Guatemala, Figure 4.2 shows how a group of indigenous people (Maya K'iché) in an urban poor community in a departmental capital of Santa Cruz del Quiché perceived and experienced poverty. A number of issues emerge from the diagram, which was drawn by the people themselves. The first is that they associate agricultural work with poverty – indigenous people in this area have historically migrated to work on plantations in the south of the country for extremely low and unstable wages (Moser and McIlwaine 2001). The diagram also illustrates that the indigenous poor do not automatically associate poverty with money and income. While 'lack of funds' and low salaries are mentioned, this relates to being unable to educate one's children and being forced to work as agricultural labourers. Instead, they associated poverty with work, few job opportunities, exploitation, discrimination, lack of education for their children, and people losing the strength to struggle for their rights. While this is not reflected in the diagram, the focus group also identified poverty as a form of violence; they felt it was unjust, and therefore violated their rights as citizens.

Although rural/urban residence, gender and ethnicity all interrelate with poverty in Middle America, other characteristics are also common among the poor. As mentioned in the examples above, lack of education is a common feature, with the poor rarely having more than a few years of primary education, and almost always having fewer years than the non-poor. This is often combined with higher levels of ill health, malnutrition

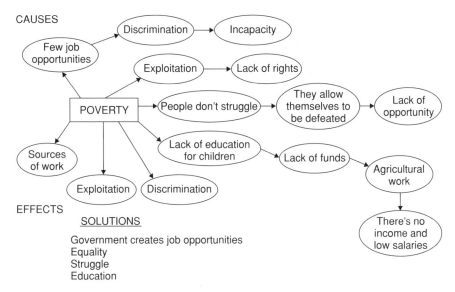

CAUSES

POVERTY

Few job opportunities

Discrimination → Incapacity

Exploitation →Lack of rights

People don't struggle → They allow themselves to be defeated → Lack of opportunity

Lack of education for children → Lack of funds

Sources of work

Exploitation   Discrimination

EFFECTS

Agricultural work

There's no income and low salaries

SOLUTIONS

Government creates job opportunities
Equality
Struggle
Education

Figure 4.2   Flow diagram of the causes, effects and solutions to poverty in Santa Cruz del Quiché, Guatemala, drawn by one woman (aged 20) and two men (aged between 40 and 60), all of Mayan origin

and infant and child mortality levels (see Figures 4.1 and 4.2). Finally, household size is often positively correlated with poverty, with the larger households more likely to be poor, and with lower dependency ratios (the proportion of workers per dependant) (see Sánchez 1997 for a summary).

Indeed, as well as being a characteristic of poverty, these factors have also been viewed as causes – by researchers and the poor themselves (Figures 4.1 and 4.2). Killick (1999) also highlights the importance of socio-political factors such as limited access to political and economic power, as well as inequalities both at the household level and within the economy as a whole. Moreover, macro-economic trends also have an influence on the causes of poverty (see below).

Poverty trends in Middle America

Bearing in mind the huge diversity in the nature of poverty in Middle America, examination of poverty trends in the region based on statistical measurements illustrates how poverty has increased since the early 1980s. Indeed, in most countries, poverty levels are higher or only slightly lower now than they were in the 1970s. For Latin America as a whole, the proportion of households living in a state of poverty was around 40 per cent in 1970, declining to 35 per cent in 1980, and rising to 41 per cent in 1990; by 1994, the proportion had only decreased to 39 per cent (Ocampo 1998: 8). In individual countries, these proportions are often much higher. In Guatemala, for example, the incidence of poverty rose from 63.4 per cent

Plate 4.2 Urban poor settlement, El Mesquital,
Guatemala City
Photo: Cathy McIlwaine

of the population living in poverty in 1981 to 79.9 per cent in 1989; extreme
poverty rose from 31.6 per cent to 59.3 per cent over the same period
(Pérez Sáinz 1998: 76). In Mexico, there were 34 per cent of households
living below the poverty line in 1984; this rose to 39 per cent in 1989, and
declined again to 36 per cent in 1992 (Sheahan 1997: 24).

Most increases in poverty have therefore occurred in the 1980s, often
referred to as the 'Lost Decade'. The only countries that did not experience
increases in poverty were the Dominican Republic (Gilbert 1997: 321) and
Panama (Ocampo 1998: 8). Moreover, urban poverty increased to a greater
extent than rural poverty during the 1980s. According to CEPAL figures,
the proportion of urban poor increased from 37 per cent to 55 per cent of
the total national population between 1970 and 1986, while the poor rural

population declined from 63 per cent to 45 per cent of the total national population (Sánchez 1997: 3). This reflects trends elsewhere in the world, where poverty has become increasingly urbanised. In addition to poverty, inequality either remained the same or increased depending on the country. This usually involved deterioration in income distribution, which has been historically unequal in the region (Altimir 1997). Even in countries where increases in poverty have not been as severe as in others, such as Panama and Costa Rica, indicators of equity and income distribution have worsened (Ocampo 1998). This deterioration has been closely linked with SAPs. Indeed, even though poverty indicators have improved in some countries in the 1990s (see for instance, Handa and King 1997 on Jamaica), it is not certain as to how sustainable this will be. Furthermore, there remain many countries, such as Nicaragua and Haiti, that are among the poorest in the world and whose prospects are not promising (see below).

Poverty and structural adjustment programmes

Adjustment refers to the neoliberal approach to development that prescribes that countries should reorient their economies along free-market lines, usually through privatisation and reducing the role of the government in the economy. As a result of the debt crisis of the 1970s, many countries in Middle America and elsewhere had little option but to turn to the international lending agencies such as the IMF and World Bank in order to keep their economies afloat. In return for reorienting their economies, adjustment loans were arranged. These usually involve initial stabilisation loans by the IMF. These are usually short-term measures, sometimes called 'austerity plans', that entail removal of state subsidies and reductions in public expenditure. Once countries have stabilised, they usually implement SAPs that are more long-term, and arranged by the World Bank. These aim to improve economic efficiency through reducing the role of the state, opening up the economy through export-led development, and deregulating markets. Some form of stabilisation and adjustment programmes has been introduced in all Middle American countries. In Latin America as a whole, 107 SAPs were implemented in 18 countries in the 1980s (Green 1995; Sparr 1996).

There has been much debate over whether SAPs have directly caused increases in poverty, especially in terms of identifying the 'counterfactual' (what would have happened if SAPs had not been implemented). Evidence still remains inconclusive. However, the key point is that it is agreed that although SAPs may not have caused poverty in a direct sense, they certainly did not lead to poverty reduction. In addition, it is now accepted that in many countries, SAPs have been associated with increasing inequalities; this is especially marked in Middle and Latin America as well as sub-Saharan Africa (Killick 1999; Simon *et al.* 1995).

More specifically, commencing in the late 1980s and continuing throughout the 1990s, there has been widespread concern about the socio-economic effects of SAPs. The most important early criticisms came from UNICEF in its publication of *Adjustment with a Human Face* (Cornia *et al.* 1987). This highlighted the social costs related to the more or less standard effects of SAPs that included increased unemployment, declining real wages, reduced public spending, especially on education and healthcare, and increased costs of basic commodities and foodstuffs. All these result directly or indirectly in increasing poverty. UNICEF also pointed out that it was women in particular who suffered. Indeed, this has become one of the major criticisms of SAPs from a gender perspective at the household level – the ways in which SAPs ignore women's contributions to domestic labour, and how it is usually women's responsibility to develop coping mechanisms when household incomes decline and prices increase (Elson 1991; Moser 1992).

These effects have been widely experienced in Middle America. Indeed, the notions of the 'structural' or 'chronic poor' and 'new poor' emerged from research in this region. In the case of Mexico, Escobar Latapí and González de la Rocha (1995) show that it was the middle classes who were the 'new poor'; during economic growth in the 1970s, they had benefited from a range of developments including expansion of the formal and public sectors, improved infrastructure and so on. With adjustment and ensuing restructuring, public-sector jobs were cut and the middle classes lost their jobs. While they point out that the structural and chronic poor were also extremely badly hit, the impact was not as severe in relative terms as that experienced by the middle classes. Similarly, there has been widespread discussion of economic restructuring and the informalisation of labour markets in the context of adjustment, both in Mexico and elsewhere in the region. In Nicaragua, for example, the first structural adjustment loan involved retrenchments of large numbers of civil servants (mainly women). While the aim was to downsize the state, no provisions were made for finding employment for those who lost their jobs. These workers therefore had to turn to subsistence informal sector activities, primarily in commerce (Pérez-Sáinz 1998; see also Chant, this volume; McIlwaine *et al.*, this volume).

Responses to poverty associated with SAPs: grassroots and policy initiatives

Dealing first with the grassroots, it is also important to mention how the poor respond in times of increased poverty, and how they are certainly not passive victims of wider economic changes. With reference to Mexico, Chant (1996b) shows how women have been disproportionately affected by recession and SAPs. However, she also discusses how resourceful women have been in dealing with poverty through creating coping strategies. Drawing on the work of González de la Rocha, and Benería and Roldán, she highlights how these responses fall into two main groups based on the Mexican experience: the first refers to curtailing household consumption, which may also be called expenditure-minimising or negative strategies. These include cutting back on the number of meals eaten, going to bed early to save electricity, and relinquishing spending on education and health services. The second approach, referred to as income-generating or positive strategies, involves women generating some form of income, sending more household members out to work, and so on. In addition to this, the poor also engage in other conscious and unconscious mechanisms such as changing household structures – often household extension – in order to bring more income-earners into the household, or to bring in people to take over domestic duties from women who need to go out to work (Chant 1996b; see also Chant, this volume).

While this reflects the resourcefulness of the poor in dealing with the exigencies affecting their daily lives, there have also been attempts to develop policies to reduce poverty at a range of levels, from community interventions to national level poverty reduction plans. It should be stressed that poverty reduction strategies are designed according to the ways in which poverty is conceived. Wratten (1995: 28) notes two opposing types of intervention that depend on the interpretations of the causes of poverty. The first is based on liberal, laissez-faire intellectual traditions and is where individuals are blamed for being unable to overcome poverty due to their own failings. The main response to this is policies linked with the free market that focus on psychological rehabilitation of the poor. The second perspective attributes the roots of poverty to unequal political and economic systems. Policy responses are associated with Marxist intellectual thought and tend to promote state intervention aimed at redistribution at all levels from the global to the household. On a more practical level, there are variations depending on whether poverty is viewed as income, or is conceived in a wider sense. Policies associated with the former invariably focus on income generation projects, such as micro-credit programmes, micro-enterprise schemes and so on. Alternatively, if poverty is viewed in a more broad-based manner, then poverty reduction projects will include education, health and nutrition elements, as well as income generation. Poverty reduction policies and programmes can be implemented by a range of actors, but are often most successful when associated with community-based organisations (CBOs) and NGOs. While they often develop their own poverty reduction projects, they also act as conduits for national-level programmes that work through them in order to implement their programmes more successfully (see Pelling, this volume, for examples that involve NGOs).

Some of the most common national-level interventions aimed at reducing poverty, especially that linked explicitly with SAPs, are Social Investment Funds. These funds were established in the late 1980s as the World Bank and other international agencies realised that poverty was increasing in the context of recession and the implementation of SAPs. Often aimed at the 'new poor' and those most vulnerable during recession, these funds have also been referred to as social safety nets, social emergency funds, or social dimensions of adjustment programmes. After the first fund in Latin America was started in Bolivia in 1986, they were created in most Central American and Caribbean countries. In the late 1980s and early 1990s, most funds involved emergency employment programmes, basic infrastructure provision, social assistance programmes and credit programmes (Siri 1996; Stahl 1996).

As a result of early criticisms of these funds, such as the fact that they are merely palliatives rather than addressing causes of poverty, and that they rarely reach the most needy and have inherent class, gender and regional biases, they have since been improved (Killick 1999). Although they were originally seen as temporary institutions (three to four years), many have been extended to become permanent. For example, in El Salvador, the *Fondo de Inversión Social* (Social Investment Fund – FIS) was established in 1990 with primary funding from the Inter-American Development Bank (IDB). The FIS remained temporary until 1997 when it was transformed into a permanent institution to deal with local development in a more integrated manner; it was renamed the FISDL – Social Investment Fund for Local Development – in an attempt to reflect its new mandate (McIlwaine 1998). In Jamaica, the traditional Social Investment Fund of the 1990s was substantially improved as a result of a participatory study of poverty and violence. On the basis of this study, the Jamaican SIF was reoriented to include more innovative projects that would build social capital in an effort to reduce poverty and violence in more sustainable ways. These included provision of conflict resolution programmes, drug-counselling facilities, training in technical skills, and family-life education and parenting courses (Moser and Holland 1997).

Other important improvements have also been made at the international policy level in relation to SAPs and dealing with poverty. As mentioned above, the World Bank, IMF and other development agencies now recognise the potentially adverse effects of SAPs. Although they will not commit to agreeing that there is a direct relationship between poverty and adjustment, the late 1990s saw important shifts towards linking macro-economic lending with poverty reduction. Therefore, as part of the Heavily Indebted Poor Country Initiative II (HIPC II) (see earlier), the IMF and World Bank are now linking lending policies with concrete poverty reduction plans. In an effort to move away from imposing conditionalities on developing countries, these organisations have instituted a series of procedures that encourage individual countries to come up with their own designs for poverty reduction in return for debt relief. For example, if a government requires debt relief (linked with previous adjustment loans that it has problems repaying), it is now expected to draft Poverty Reduction Strategy Papers (PRSPs) in consultation with civil society covering poverty issues and policies over a three-year timeframe, together with how this will be budgeted. Moreover, these PRSPs are also encouraged to use participatory assessment techniques in order to discover how the poor really experience poverty (Healey *et al.* 2000).

Although it is still in its infancy, there have been criticisms of this new poverty-focused approach. In the case of Honduras, which has a new Poverty Reduction Growth Facility (PRGF) lending programme from the IMF, Oxfam (2000) argues that this is little different from structural adjustment loans. In addition, the World Bank is preparing to invest in a social fund that does not address the underlying causes of poverty (ibid.). Moreover, in the case of Middle America, Oxfam points out that many countries that suffer from acute poverty and debt problems, such as Haiti and Jamaica, are excluded inexplicably from the new initiative. Therefore, while these are encouraging developments, many challenges remain (Healey *et al.* 2000). The example of Nicaragua, Middle America's poorest and most indebted nation, highlights these challenges (see Box 4.3).

Box 4.3   Poverty reduction in Nicaragua:
adjustment and the HIPC Initiative

Nicaragua is one of the poorest countries in Middle America, with a range of
social indicators reflecting extreme poverty. For example, more than half the
population live below the poverty line and two-fifths of poor children are
malnourished. Almost two-fifths of the population have no access to safe
drinking water and three-quarters are without sanitation facilities. On top of
this, the country has the world's highest level of per capita debt – US$1,3000
per capita. Therefore, in 1997 Nicaragua spent two-and-a-half times health and
education spending on debt repayments, and 11 times spending on primary
healthcare.

On the basis of a series of indicators, Nicaragua is eligible for debt relief
under the HIPC initiative. For example, the debt service ratio is 51 per cent for
debt service due, compared with the HIPC eligibility range of 20–25 per cent.
Yet, although the country was identified as eligible in 1998, it will not receive
any relief before 2002. This is because of a series of conditionalities that the
country must first adhere to. These include undergoing two successive IMF
programmes lasting a total of six years before debt relief can be granted – this
period can only be reduced if the Enhanced Structural Adjustment Facility
(ESAF) has been strong enough. This has not been the case in Nicaragua.
Furthermore, many of the delays with the adjustment loans were the result of
disagreements between the government and the IMF on procedural issues,
while the poor suffer.

Therefore, for Nicaragua, although the HIPC initiative must be welcomed,
the lengthy and conditional process could potentially threaten to push even
more Nicaraguans below the poverty line.

Source: Oxfam 1998.

Conclusion

Poverty in Middle America is multifaceted and means more than income
poverty alone. Instead, it encompasses a range of subjective aspects that
can be identified and understood through the use of such terms as entitlements, vulnerability and social exclusion, and examined using participatory
appraisal methodologies. As well as adopting a wide-ranging interpretation
of poverty, it is also important to recognise the diversity of who is poor,
especially on grounds of gender, race and ethnicity, as the way in which
men and women from different ethnic groups experience poverty differs
depending on context and culture. In terms of the dynamics of poverty,
various forms of deprivation remain of major concern in the region, especially in countries such as Nicaragua, Haiti and Guatemala. Indeed, it is
important to stress that income poverty levels are worse in Middle America
than in Latin America as a whole. Poverty has also been closely associated

with SAPs and economic restructuring in the region. Regardless of the intricacies of the relationships between SAPs and poverty, poverty indicators worsened almost everywhere during the 1980s. Although there have been some improvements in the 1990s, it remains to be seen if these will be sustainable into the twenty-first century.

Useful websites

www.oneworld.net A development website with links to an extremely wide range of poverty-related sites, including many NGOs and development agencies.

www.odi.org.uk/briefing Includes 'Poverty briefings' – up-to-date issues relating to poverty and development.

www.oxfam.org/what_does/advocacy/papers/htm A series of papers prepared by Oxfam on poverty and debt issues.

www.worldbank.org/poverty World Bank information on poverty including trends, data and research.

www.ids.ac.uk/bridge Bridge Reports dealing with gender and poverty issues.

Further reading

Allen, T. and **Thomas, A.** (eds) (2000) *Poverty and Development into the 21st Century.* Oxford University Press, Oxford.
A clear set of chapters outlining various dimensions of poverty and development, conceptually and including empirical examples.

Baden, S. with **Milward, K.** (1995) *Gender and Poverty.* Bridge Report No. 30, Institute of Development Studies, Brighton, Sussex.
A good summary of the conceptual issues in relation to gender and poverty covering female-headed households and women's employment experiences.

Gilbert, A. (1997) Poverty and social policy in Latin America, *Social Policy and Administration* 31(4): 320–335.
An article full of useful facts and figures, as well as an outline of the issues relating to the nature of poverty in Latin America.

Tardanico, R. and **Menjívar Larín, R.** (eds) (1997) *Global Restructuring, Employment, and Social Inequality in Urban Latin America.* North–South Center Press, University of Miami, Coral Gables.
A useful edited collection that includes chapters on a number of Middle American countries including the Dominican Republic, Guatemala, Mexico and Costa Rica.

Wratten, E. (1995) Conceptualizing urban poverty, *Environment and Urbanization* 7(1): 11–36.

From a special issue on urban poverty, this article outlines the conceptual approaches to poverty from poverty lines to the subjective, participatory perspectives.

References

Altimir, O. (1994) Distribución del ingreso e incidencia de la pobreza a lo largo del ajuste, *Revista de CEPAL* **52**: 7–32.

Altimir, O. (1997) Desigualdad, empleo y pobreza en America Latina: efectos del ajuste y del cambio en el estilo de desarrollo, *Desarrollo Económico* **37**(145): 3–30.

Amis, P. and **Rakodi, C.** (1994) Urban poverty: issues for research and policy, *Journal of International Development* **6**(5): 627–634.

Baden, S. with **Milward, K.** (1995) *Gender and Poverty.* Bridge Report No. 30, Institute of Development Studies, Brighton.

Baulch, B. (1996) The new poverty agenda: a disputed consensus, *IDS Bulletin* **27**(1): 1–10.

Bebbington, A. (1999) Capitals and capabilities: a framework for analysing peasant viability, rural livelihoods and poverty, *World Development* **27**(12): 2021–2044.

Chambers, R. (1995) Poverty and livelihoods: whose reality counts?, *Environment and Urbanization* **7**(1): 173–204.

—— (1997) Editorial: responsible well-being – a personal agenda for development, *World Development* **25**(11): 1743–1754.

Chant, S. (1996a) *Gender, Urbanization and Housing: Issues and Challenges for the 1990s.* UNDP, New York.

—— (1996b) Women's roles in recession and economic restructuring in Mexico and the Philippines, *Geoforum* **27**(3): 297–327.

—— (1997a) *Women-Headed Households: Diversity and Dynamics in the Developing World.* Macmillan, Basingstoke.

—— (1997b) Women-headed households: poorest of the poor? Perspectives from Mexico, Costa Rica and the Philippines, *IDS Bulletin* **28**(3): 26–47.

Chant, S. and **McIlwaine, C.** (1998) *Three Generations, Two Genders, One World: Women and Men in a Changing Century.* Zed, London.

Cornia, G.A., Jolly, R. and **Stewart, F.** (1987) *Adjustment with a Human Face: Protecting the Vulnerable and Promoting Growth. A Study by UNICEF.* Clarendon Press, Oxford.

De Haan, A. (1998) 'Social exclusion': an alternative concept for the study of deprivation, *IDS Bulletin* **29**(1): 10–19.

De Haan, A. and **Maxwell, S.** (1998) Poverty and social exclusion in the North and South, *IDS Bulletin* **29**(1): 1–9.

Elson, D. (1991) Structural adjustment: its effect on women, in Wallace, T. with March, C. (eds) *Changing Perceptions*. Oxfam, Oxford: 39–53.

Escobar Latapí, A. and **González de la Rocha, M.** (1995) Crisis, restructuring and urban poverty in Mexico, *Environment and Urbanization* **7**(1): 57–76.

Fine, B. (1999) The development state is dead – long live social capital?, *Development and Change* **30**: 1–19.

Fuwa, N. (2000) The poverty and heterogeneity among female-headed households revisited: the case of Panama, *World Development* **28**(8): 1515–1542.

Gafar, J. (1998) Growth, inequality and poverty in selected Caribbean and Latin American countries, with emphasis on Guyana, *Journal of Latin American Studies* **30**: 591–617.

Gilbert, A. (1997) Poverty and social policy in Latin America, *Social Policy and Administration* **31**(4): 320–335.

Green, D. (1995) *The Silent Revolution: the Rise of Market Economics in Latin America*. LAB, London.

Handa, S. and **King, D.** (1997) Structural adjustment policies, income distribution and poverty: a review of the Jamaican experience, *World Development* **25**(6): 915–930.

Harriss, J. and **De Renzio, P.** (1997) An introductory bibliographic essay. 'Missing link' or analytically missing? The concept of social capital, *Journal of International Development* **9**(7): 919–937.

Healey, J., Foster, M., Norton, A. and **Booth, D.** (2000) *Towards National Public Expenditure Strategies for Poverty Reduction*. ODI Poverty Briefing, Overseas Development Institute, London (*www.odi.org.uk/briefing/ pov7.html*).

International Institute for Labour Studies/United Nations Development Programme (IILS/UNDP) (1996) *Social Exclusion and Anti-Poverty Strategies*. IILS, Geneva.

Jackson, C. (1996) Rescuing gender from the poverty trap, *World Development* **24**(3): 489–504.

Kabeer, N. (1996) Agency, well-being and inequality: reflections on the gender dimensions of poverty, *IDS Bulletin* **27**(1): 11–21.

—— (1997) Editorial: tactics and trade-offs: revisiting the links between gender and poverty, *IDS Bulletin* **28**(3): 1–13.

Killick, T. (1999) *Making Adjustment Work for the Poor*. ODI Poverty Briefing, Overseas Development Institute, London (*www.odi.org.uk/briefing/ pov5.html*).

Krishna, A. and **Uphoff, N.** (1999) *Mapping and Measuring Social Capital: a Conceptual and Empirical Study of Collective Action for Conserving and Developing Watersheds in Rajasthan, India.* Social Capital Initiative Working Paper No. 13, World Bank, Washington, D.C.

Maxwell, S. (1999) *The Meaning and Measurement of Poverty.* ODI Poverty Briefing, Overseas Development Institute, London (*www.odi.org.uk/ briefing/pov3.html*).

McIlwaine, C. (1997) Vulnerable or poor? A study of ethnic and gender disadvantage among Afro-Caribbeans in Limón, Costa Rica, *The European Journal of Development Research* **9**(2): 35–61.

—— (1998) Contesting civil society: reflections from El Salvador, *Third World Quarterly* **19**(4): 651–672.

Menjívar, R. and **Feliciani, F.** (eds) (1995) *Análisis de la Exclusión Social a Nivel Departamental: Los Casos de Costa Rica, El Salvador y Guatemala.* FLACSO, PNUD, UNOPS, PRODERE-Edinfodoc, Guatemala City.

Moser, C. (1992) Adjustment from below: low-income women, time and the triple role in Guayaquil, Ecuador, in Afshar, H. and Dennis, C. (eds) *Women and Adjustment Policies in the Third World.* Macmillan, Basingstoke: 87–116.

—— (1998) The asset vulnerability framework: reassessing urban poverty reduction strategies, *World Development* 26: 1–19.

Moser, C. and **Holland, J.** (1997) *Urban Poverty and Violence in Jamaica.* World Bank, Washington D.C.

Moser, C. and **McIlwaine, C.** (1999) Participatory Urban Appraisal and its application for research on violence, *Environment and Urbanization* **11**(2): 203–226.

—— (2000) *Urban Poor Perceptions of Violence and Exclusion in Colombia,* World Bank, Washington, D.C.

—— (2001) *Violence in a Post-Conflict Context: Urban Poor Perceptions from Guatemala.* World Bank, Washington, D.C.

Ocampo, J.A. (1998) Income distribution, poverty and social expenditure in Latin America, *CEPAL Review* **65**: 7–14.

Ottone, E. (1997) Overcoming poverty and exclusion as causes of insecurity in Latin America, *Security Dialogue* **28**(1): 7–16.

Oxfam (1998) Debt Relief for Nicaragua: Breaking Out of the Poverty Trap. Policy Paper, Oxfam, Oxford (*www.oxfam.org/advocacy/papers/ nicaragua.html*).

—— (2000) Media Briefing: Debt Relief and Poverty Reduction: Failing to Deliver. Oxfam, Oxford (*www.oxfam.org/advocacy/papers/htm*).

Pelling, M. (1997) What determines vulnerability to floods: a case study in Georgetown, Guyana, *Environment and Urbanization* **9**(1): 203–226.

Pérez Sáinz, J.P. (1998) The new faces of informality in Central America, *Journal of Latin American Studies* **30**: 157–179.

Pinker, R. (1999) Do poverty definitions matter? in Gordon, D. and Spicker, P. (eds) *The International Glossary on Poverty*. Zed, London: 1–5.

Portes, A. (1998) Social capital: its origins and applications in modern Sociology *American Review of Sociology* **24**: 1–24.

Programa de Naciónes Unidas de Desarrollo (PNUD – UNDP) (1998) *Guatemala: Los Contrastes del Desarrollo Humano*. UNDP, Guatemala.

Rahnema, R. (1995) Poverty, in Sachs, W. (ed.) *The Development Dictionary: a Guide to Knowledge as Power*. Zed, London: 158–176.

Recovery of Historical Memory Project (REHMI) (1999) *Guatemala: Never Again*. CIIR and LAB, London.

Robb, C. (1999) *Can the Poor Influence Policy? Participatory Poverty Assessments in the Developing World*. World Bank, Washington D.C.

Rubio, M. (1997) Perverse social capital: some evidence from Colombia, *Journal of Economic Issues* **31**(3): 805–816.

Sánchez, D. (1997) *Poverty in Latin America*. International Development Research Centre (IDRC), Ottowa (*www.idrc.ca/lacro/publicaciones*).

Sen, A. (1981) *Poverty and Famines: an Essay on Entitlement and Deprivation*. Clarendon Press, Oxford.

Sheahan, J. (1997) Effects of liberalization programs on poverty and inequality, *Latin American Research Review* **32**(3): 7–37.

Simon, D., van Spengen, W., Dixon, C. and **Närman, A.** (1995) (eds) *Structurally Adjusted Africa: Poverty, Debt and Basic Needs*. Pluto, London.

Siri, G. (1996) Social Investment Funds in Latin America, *CEPAL Review* **59**: 73–82.

Sparr, P. (1996) (ed.) *Mortgaging Women's Lives: Feminist Critiques of Structural Adjustment*. Zed, London.

Stahl, Karin (1996) Anti-poverty programmes: making structural adjustment more palatable, *NACLA Report on the Americas* xxix: 32–6.

Thomas, T. (2000) Poverty and the 'end of development', in Allen, T. and Thomas, A. (eds) *Poverty and Development into the 21st Century*. Oxford University Press, Oxford: 3–22.

Trotz, A. (1996) Gender, ethnicity and familial ideology in Georgetown, Guyana: household structure and female labour force participation reconsidered, *European Journal of Development Research* **8**(1): 177–199.

United Nations Centre for Human Setflements (UNCHS) (1996) *An Urbanizing World: Global Report on Human Settlements 1996*. Oxford University Press, Oxford and New York.

United Nations Development Programme (UNDP) (1990) *Human Development Report 1990*. Oxford University Press, Oxford and New York.

—— (1995) *Human Development Report 1995*. Oxford University Press, Oxford and New York.

—— (1996) *Human Development Report 1996*. Oxford University Press, Oxford and New York.

—— (1999) *Human Development Report 1999*. Oxford University Press, Oxford and New York.

UNICEF/Oxfam (1999) *Debt Relief and Poverty Reduction: Meeting the Challenge*. Oxfam, Oxford (*www.oxfam.org/advocacy/debtchallenge.htm*).

Watkins, K. (1995) *The Oxfam Poverty Report*. Oxfam, Oxford.

Webster, A. (1990) *Introduction to the Sociology of Development* (2nd edn.). Macmillan, Basingstoke.

World Bank (1990) *World Development Report 1990*. Oxford University Press, Oxford and New York.

—— (2000) *World Development Report 2000/2001*. Oxford University Press, Oxford and New York.

Wratten, E. (1995) Conceptualizing urban poverty, *Environment and Urbanization* **7**(11): 11–36.

CHAPTER 5

Making a living: employment, livelihoods and the informal sector[1]

Cathy McIlwaine, Sylvia Chant and Sally Lloyd Evans

Introduction

This chapter explores the ways in which people in Middle America make a living. By emphasising the notion of livelihoods, it shows how much of the work undertaken in the region takes place in the informal sector, or increasingly, under informalised conditions within the formal labour market. Indeed, a major aim of the chapter is to illustrate how processes of globalisation, especially those linked with recession and structural adjustment programmes (SAPs), have led to marked reconfigurations of the labour markets of the region.

The main focus in the chapter is on urban areas because, in the light of the decline in agricultural employment since the second half of the twentieth century, labour markets in the region are now predominantly urban (Chant 1999a). In addition, the chapter focuses on the informal sector, given that formal sector employment is discussed elsewhere in this volume in relation to export-oriented factory employment (Willis), ecotourism (Barton) and agriculture (Thorpe and Bennett). However, the linkages between the two sectors, and the contemporary difficulties in treating the two sectors as discrete and separate, are recognised throughout. In turn, the chapter also seeks to disaggregate the informal sector according to gender, ethnicity and age, showing that there are particular concentrations of groups in informal employment. A final major issue addressed is the role of the informal sector as a mechanism of survival versus a 'sector of entrepreneurship', and the question of whether policies should encourage or inhibit its growth.

[1] The phrase 'making a living' has also been used elsewhere in chapter titles (see for example, Wield and Chataway 2000). However, we felt that this was the most appropriate phrase to capture the nature of employment, especially informal work, in Middle America.

Employment and livelihoods in Middle America: an overview

Throughout the twentieth century, the concepts of 'employment' and 'work' were subject to much debate. This was heightened as attempts were made to transfer Western interpretations of employment and work to the developing world. For instance, the term 'employment' generally refers to work carried out in return for a wage or other form of remuneration. However, measurement of employment usually includes only that registered in national statistics, thereby excluding forms of employment undertaken in the informal economy or the household. 'Work' on the other hand, is a broader term that includes all productive activities, including those that are unwaged and not directly remunerated. Thus, household reproduction, community activities, and unpaid subsistence production (for the producer's own use), often carried out by women in the home, are incorporated here (Potter and Lloyd Evans 1998: 160). This wider conceptualisation is particularly important in the developing world where livelihoods and making a living are often ensured through unwaged and subsistence activities rather than waged employment in a strict sense (Chant 1999b; Wield and Chataway 2000).

Similarly, the issue of unemployment is also problematic in that it is generally associated with the wage economy (Gilbert and Gugler 1992). Officially, unemployment refers to the situation where someone is without paid work or self-employment, but is currently looking for work, and is therefore part of the 'economically-active population' (EAP). Indeed, research sometimes referred to as the 'luxury unemployment thesis' has shown that only the educated, relatively well off, and the young starting out on their careers are able to be unemployed in the absence of social security and/or assistance from family (Potter and Lloyd Evans 1998: 169). In the light of the problems associated with this essentially ethnocentric concept, the notion of 'underemployment', sometimes known as 'disguised unemployment', is often perceived to be more expedient. This denotes situations where people work fewer hours than they would like, where they work part-time or occasionally, where their skills are underutilised, or where productivity is low (Gilbert and Gugler 1992).

In view of these limitations, it is more appropriate to discuss employment in Middle America, as in the rest of the developing world, in relation to livelihoods and the notion of 'making a living' (Wield and Chataway 2000). This frequently involves the self-styled generation of means to make an income, or the development of strategies to ensure survival that perhaps entails larger collectivities such as households or kin networks. Given the difficulties of obtaining wage employment in the formal sector, it is no surprise that in both rural and urban areas a large proportion of the economically active population work outside the formal economy. In some Middle American countries, such as Nicaragua, almost two-thirds of the labour force works outside the waged sector (Pérez Sáinz 1998). Therefore, when exploring the nature of employment, and especially urban employment, it is important to adopt a holistic approach that incorporates notions of 'work', 'underemployment', and 'livelihoods'. Above all, consideration of the 'informal sector' and the nature of the increasing informalisation of labour markets is crucial in understanding urban economies in Middle America.

Definitions and characteristics of formal and informal sectors

Consideration of the concepts of formal and informal sectors is essential when examining the nature of urban labour markets in Middle America and elsewhere, notwithstanding that the two are closely interlinked (see below). In broad terms, however, the formal sector denotes waged, protected and regulated employment, covering all branches of the economy from manufacturing and services to the public sector. Depending on the degree of regulation, formal employment can range from large-scale capital-intensive factory production using high-technology inputs, to small service-oriented businesses providing contracts, fixed wages and benefits.

The relative ease in defining the formal sector is in direct contrast to the informal sector. The concept of the 'informal sector' has been interpreted in a multitude of ways since it was first developed in the 1970s. Indeed, until the 1980s it was usually characterised in a dichotomous manner as the opposite to the formal sector (see below for further discussion). Generally, the informal sector refers to a range of casual and irregular work that includes petty trading, self-employment, work in personal services or in small micro-enterprises (Bromley 1979; Wield and Chataway 2000: 107). Possibly its most significant defining feature is that it is largely unregulated in respect of registration of a business, the payment of taxes, and compliance with legislation on working hours, social security contributions, wages and fringe benefits (see Tokman 1991: 143). Although non-compliance with various of these regulations is also found in the formal sector (Roberts 1994: 16; see also later), the irregularity and invisibility of the informal sector in national accounting data means that it is sometimes referred to as 'precarious' or 'subterranean' employment (Portes and Schauffler 1993: 33). Terminology also differs from country to country, as well as within countries. In Trinidad, for example, the informal sector is variously referred to as the 'subsistence sector', 'underground sector', 'invisible sector', 'underground economy' and the 'black economy' (Lloyd Evans 1998).

Derived from the research of anthropologist Keith Hart in Ghana, funded by the International Labour Office (ILO) in the early 1970s, the term 'informal sector' was developed in the context of large-scale rural–urban migration, accompanied by growing levels of urbanisation and limited labour demand in the emerging industrial sectors of developing economies. In recognising that there was insufficient job provision in the formal sector, Hart and others discovered that people created a variety of mechanisms to generate income and make a living in what was termed the 'informal sector'. Initially distinguishing between salaried jobs in the formal sector and self-employment in the informal sector, Hart further subdivided the informal economy into 'legitimate' and 'illegitimate' activities. While 'legitimate' jobs such as small-scale commerce, personal services and home-based production made a contribution to economic growth, 'illegitimate' activities included prostitution, begging, stealing and scavenging• (Hart 1973). Although the existence of the informal sector had been documented since the 1950s in countries such as Jamaica, Mexico and Puerto Rico (Potter and Lloyd Evans 1998: 174), it was the interest generated by the ILO's so-called 'Kenya Mission' that led to widespread recognition of this previously ignored sector of urban economies in the South.

Bearing in mind context-specific variations, the informal sector generally provides opportunities to make a living for many low-income groups in Middle America. While there are some common characteristics such as those outlined above, the activities included within the informal sector are highly heterogeneous (Chant 1999a). Activities may range from unpaid work in the home in family-run manufacturing workshops or selling home-cooked food or ice and soft drinks in 'front room' eateries, to more organised occupations that are linked with the formal economy, such as various types of subcontracting common in the apparel and footwear industries (see Benería and Roldán 1987; Chant 1991 on Mexico). Many informal activities have local names and meanings; for example, in Santo Domingo in the Dominican Republic, mobile retail vendors who sell fruit and vegetables from tricycles in the street are called *lechugueros* (literally 'lettuce people') (Itzigsohn 1997: 68). In a study in Trinidad, where an informal activity is defined by the Central Statistics Office as one operating under the evasion of at least one government law or regulation, Lloyd Evans (1998) outlines a typology of four different types of activities: 1) retailing, 2) small-scale production, 3) services, and 4) goods and services produced within the formal sector by informal means. Within these broad categories, a total of 75 different operations were identified, highlighting the rich diversity of informal sector occupations within a given country.

Turning to the relative size of the informal sector, it is important to bear in mind that data are unreliable in the light of the irregular and underground nature of many of its activities. Indeed, most studies of the informal sector are based on specific studies rather than official figures that are difficult to come by. Even when official statistics exist, definitions vary from country to country (Tardanico 1997). Acknowledging this, the informal sector in Middle American countries is relatively large, often constituting more than half of all workers. Based on household surveys using a standard definition of informal activities as establishments with less than five employees, Pérez Sáinz (1998: 131) reports levels of informal sector employment in non-agricultural employment between 1989 and 1993 as 53 per cent in Guatemala, 55.3 per cent in El Salvador, 48.9 per cent in Honduras, and 63.3 per cent in Nicaragua.

More specifically, Table 5.1 illustrates that in four selected countries in the region, the majority of informal activities relate to services. In Mexico, production was evenly distributed between transport and services. In terms of the gender of the workers involved in the informal sector, generally

Plate 5.1 Informal sector shoe-making enterprise (father and his two sons), Esquipulas, Guatemala
Photo: Cathy McIlwaine

Table 5.1 Percentage of production which is informal in four Middle American countries

| | *Percentage of production which is informal* | | | |
	Manufacturing	*Transport*	*Services*	*Total*
Costa Rica (1984)	14	9	16	15
Honduras (1990)	26	17	28	26
Jamaica (1988)	19	23	30	25
Mexico (1992)	9	20	20	16

Source: United Nations (1995: 135).

women outnumber men, although this is not always the case. For instance, while in Costa Rica, Honduras and Jamaica, more women than men are employed in informal employment, in Mexico more men than women earn a living in the informal economy (Table 5.2).

Table 5.2 Male and female labour force in different components of the informal sector in four selected Middle American countries

	Percentage of labour force which is informal							
	Manufacturing		*Transport*		*Services*		*Total*	
	Men	*Women*	*Men*	*Women*	*Men*	*Women*	*Men*	*Women*
Costa Rica (1984)	14	13	11	0	7	22	8	19
Honduras (1990)	15	52	29	0	26	29	21	34
Jamaica (1988)	21	11	29	0	27	32	25	28
Mexico (1992)	8	11	21	2	30	16	22	15

Source: United Nations (1995: 136).

Trends in employment and livelihoods

Bearing these definitional issues in mind, patterns of employment and livelihoods in the last two decades of the twentieth century in Middle America, and in Latin America generally, have been subject to widespread change. On top of the progressive urbanisation of employment from the 1980s onwards, debt, recession, SAPs and globalisation steered Middle American economies along increasingly liberalised and export-oriented lines. Aside from the increased domination by multinational corporations of Middle American labour markets (Ward and Pyle 1995: 38), one major change has been the increasing informalisation of labour markets, not only in respect of the size of the informal sector, but also in terms of conditions for workers within the formal economy (Chant 1999a).

Changes in the formal sector in the context of adjustment

Rates of formal employment in Middle America have changed over time, especially between the 1970s and the present day. Considering unemployment (which relates mainly to formal economies where measurements can be made), levels in Latin America as a whole rose from 6 per cent in 1974 to 14 per cent in 1984 (Cubitt 1995: 164). Although levels had dropped back to between 4 and 7 per cent in most parts of the region by the early 1990s, in some Middle American countries unemployment increased dramatically during this time. In Nicaragua, for instance, 23.5 per cent of the labour force was unemployed in 1994 compared with 4.5 per cent in 1986 (Bulmer-Thomas 1996: 326). Unemployment rates tend to be highest among youthful populations, with highest increases noted among young people aged 15 to 24 who have between 6 and 12 years of schooling (i.e. not the least educated) (UN/ECLAC 1994: 31). In Panama, for example, open unemployment for all economically active people was 18.6 per cent in 1991, but the level for 15- to 24-year-olds was 35.1 per cent (ibid.: 144). In Mexico, in 1998, similar patterns can be observed. While unemployment

115

among women stood at 3.8 per cent this year, and men's at 2.5 per cent, for 15- to 24-year-olds the corresponding figures were 6.4 per cent and 4.7 per cent (UNDP 2000: 260). These situations partly reflect the population structure as well as increased economic participation of women (ibid.: 31). In terms of underemployment, levels grew by 48 per cent between 1980 and 1985 in Latin America as a whole (Safa 1995: 33).

Other changes have occurred in relation to the make-up of the formal sector. First, diversification of branches of employment, with a shift towards export-oriented activities linked with export manufacturing and international tourism (see Barton, Willis, this volume); and second, a downsizing of the public sector as part of state reform and a reduction in public expenditure. Associated with economic recession and SAPs, many governments in the region were forced to reduce state machinery and retrench public sector workers. In addition, many organisations linked with the state were privatised, often involving widespread job losses (Chant 1999a). In the Nicaraguan case, at the beginning of the Chamorro regime in 1990, the first SAP involved the retrenchment of 250,000 public sector employees (Green 1995: 56–7). Although a conversion programme was established to retrain former public sector employees (known as *Reconversión Ocupacional* – Occupational Conversion), demand greatly exceeded the number of places on the scheme. Furthermore, it was women, and especially mature women, who were disproportionately affected by these processes. These former civil servants invariably created livelihood opportunities in the informal economy, with most establishing small commercial establishments in their homes. However, these barely covered costs, and were never more than subsistence activities (Pérez Sáinz 1998) (see below).

Another area of marked change in the way in which the formal sector functioned in Middle America relates to working conditions and wages. Responses to economic recession and neoliberal restructuring have involved an increased incidence of subcontracting and short-term contracts, especially in manufacturing. Trade union activities have also been restricted, as policies have given greater powers to employers to dismiss staff and hire them on a temporary basis (Green 1996: 109–10). In Mexico, shoe manufacturers in the city of León have subcontracted out increasing amounts of production to small-scale home-based workshops and to individual outworkers to make production more flexible and to cut labour costs (Chant 1991). These shifts have also entailed widespread erosion of worker protection from labour laws, which were often inadequate in the first place (Thomas 1996). Integral to these cost-cutting exercises has been the decline of real wages. Due to inflationary costs of living, there have been negative growth rates in average real earnings in many countries in Middle America between 1970 and the mid-1990s. For instance, in Guatemala annual growth rates in real earnings were –3.2 per cent between 1970 and 1980, and –1.6 per cent between 1980 and 1992 (Chant 1999a: 254). The minimum wage also declined substantially in real terms in the 1980s (Thomas 1996: 90–1). In Mexico, for example, the real minimum urban wage was only 42 per cent of its 1980 value in 1994 (Thomas 1999: 270n).

Evolution of the informal sector in the context of adjustment

These changes in formal sector employment are directly related to the evolution of the informal sector. Throughout Middle America, the proportion of the economically active labour force involved in informal sector activities has increased over time. Estimates for Latin America as a whole suggest that the share of the workforce in informal activities rose from 16.9 to 19.3 per cent between 1970 and 1980 (Tokman 1989: 1067). Between 1980 and 1990, levels of informal employment grew further with the proportion of workers in the informal labour force in cities rising from an overall average of 25.6 per cent to 30 per cent; this was considerably more than growth in the formal sector (Gilbert 1994). Indeed, it has been argued that between 1980 and 1993, 82 out of every 100 new jobs created in Latin America more generally were in the informal sector (Potter and Lloyd Evans 1998).

In terms of explanations for these patterns, increases during the 1960s and 1970s are usually attributed to labour surpluses in cities arising from rural–urban migration (Portes and Schauffler 1993). In the 1980s, however, informal sector growth was more closely linked with the consequences of recession, restructuring and SAPs (Itzigsohn 1997 on the Dominican Republic). The changes in the formal sector associated with reductions in public sector employment, increased subcontracting by manufacturing firms, as well as an overall decline in demand for formal sector workers, have led to concomitant increases in informal activities (Portes and Itzigsohn 1997). As mentioned earlier with reference to Nicaragua, informal activities are often the only types of livelihood strategies that former public sector workers can turn to in the face of unemployment. In addition, some smaller firms may become informal in light of recession as they become less able to pay registration, as well as tax and labour overheads (Escobar 1988). Linked with this is that the middle classes, as well as the poor, are increasingly forced to turn to informal sector activities in order to protect incomes and consumption. With declining wages and increasing prices, middle-class people have been forced to engage in multiple income-generating activities, often in conjunction with their formal daytime jobs (Lozano 1997 on the Dominican Republic). For instance, in countries such as the Dominican Republic where the decline in the real minimum wage affected middle-class public sector workers, many people were forced to establish small-scale businesses such as taking in lodgers, taking on tutoring jobs, or even becoming a taxi-driver in the evenings. In turn, the formal sector itself is becoming increasingly informalised, exemplified by the shift towards subcontracting work and deregularisation (Standing 1999; see also above).

Although the informal sector provides essential livelihood opportunities for many people, and sometimes allows for the development of highly profitable enterprises (see below), working conditions have become increasingly competitive during the last two decades. People have not only diversified their activities, but have also intensified them (Moser 1998). This often involves engaging in more than one job at a time (see above), as well as more hours per day. Furthermore, the market for goods sold in the informal sector is narrowing as more and more people are providing the same products and services. In Mexico, for example, Escobar Latapí and González de la Rocha (1995) noted that the self-employed in the late 1980s and early 1990s, such as mechanics, cobblers, joiners and market sellers, worked considerably longer hours than their counterparts in the 1960s in an effort to make ends meet. This is exacerbated by the lower purchasing power among the population in general and greater labour demand as populations expand. These factors in combination have led to a decrease in wealth generated within the informal sector; according to ILO figures for Latin America and the Caribbean, there was a 42 per cent relative drop in incomes between 1980 and 1989 (Moghadam 1995: 122–3). As such, there are serious limits to the prospective expansion of the informal sector.

As with formal sector retrenchments, certain groups are more vulnerable than others. In many Middle American countries, women suffer most, given their disproportionate concentration in the sector and because they tend to be engaged in activities that require few skills yet are highly competitive (Bromley 1997). In Costa Rica, for example, where 41 per cent of the informal workforce is female, low-income women in the northwest province of Guanacaste complain that their limited skills and capital resources restrict them to casual ventures such as selling home-made sweets, flavoured ices and pastries outside local schools or on the streets (Chant 1999b). These can often lead to the 'discouraged worker' effect where people give up attempting to establish an income-generating activity in the knowledge that making a living will be so difficult (Baden 1993: 13). Other adverse effects of adjustment also fall heavily on the shoulders of women, such as the lifting of food subsidies, and greater influxes of imported convenience foods stemming from trade and currency liberalisation (Chant 1996; Tinker 1997).

Formal and informal sector interrelations: a continuum of activities

The preceding discussion of changes in formal and informal employment over time, especially in the context of economic recession and restructuring linked with SAPs, has suggested that the two sectors are interrelated rather than separate entities within urban labour markets. On the one hand, the formal sector is becoming increasingly informalised, and on the other, detailed empirical studies from Middle America and elsewhere have shown that the informal sector is linked with the formal sector in a range of subordinate ways (see Bromley 1997 on an update of his classic study of Cali, Colombia). In terms of the latter, the formal sector has always drawn on informal activities to produce goods, services and markets, as well as distribution (Roberts 1995).

These studies have challenged the notion of labour market dualism that was common in accounts of labour markets in the 1970s. Labour market dualism assumes that the two sectors of the economy – formal and informal – are autonomous and separate entities. In contrast, and drawing on neo-Marxian theories on 'petty commodity production', Moser (1978) suggested that the labour market is more usefully conceptualised as a continuum of productive activities. This continuum involved a range of levels and variations in formality and informality, with few enterprises or activities being completely formal or entirely informal. Furthermore, this neo-Marxian perspective also argued that the relationships among different types of employment were exploitative rather than benign (see Thomas 1995). More recently, the concept of a continuum has been worked into the thesis of 'structuralist articulation' that perceives urban labour markets as integrated systems based on complex networks of relations among different formal and informal businesses and activities (Portes and Schauffler 1993: 48).

It should be emphasised that although some of these links may be detrimental to informal firms, in some cases, they are also essential for informal enterprises to flourish (Portes and Itzigsohn 1997: 240–1). Indeed, in a limited number of cases globalisation processes through transnational corporation activities and the diversification of subcontracting arrangements have been extremely beneficial for informal firms (Portes and Landolt 2000). Often there are layers of different activities that provide employment for a wide range of people. In some cases, these linkages are also spatially distributed both within national territories and internationally, highlighting the importance of globalisation processes. In an example of clothing manufacture among Mayan micro-entrepreneurs in Guatemala, Pérez Sáinz (1997a) outlines a complex system of interrelationships operating nationally and internationally. These micro-entrepreneurs produce clothing in small workshops (in a small urban centre 25 kilometres from the capital, Guatemala City). They are inserted into the subcontracting process in three main ways. First, some enterprises produce clothing directly for foreign (North American) firms; second, some businesses supply national garment-exporting firms; and third, some workshops produce for national businesses located in Guatemala City that sell the goods locally. While a relatively high degree of prosperity was noted among these artisans, they were also subordinate and highly vulnerable to changes in transnational policies given the 'top-down' system of production (see also Box 5.1 on an example from Jamaica).

Gender, ethnicity and the informal sector

In understanding the nature and functioning of the informal sector in Middle America, it is essential to recognise gender and ethnic variations in its types of activities and operations. Dealing first with gender, as already indicated in Table 5.2, a large proportion of informal activities are carried out by women, although this is not always the case. Indeed, Scott (1995) points out that although women are disproportionately employed in the informal sector, they do not dominate it when it is examined as a whole (with the exception of domestic service). None the less, the phenomenon of the 'global feminisation of labour' (Standing 1989) is associated with

---

Box 5.1 Interrelations between formal and informal sectors: the Jamaican ackee industry

The fruit packing and processing industry, and more specifically the canning and export of ackee fruit which is a national dish in Jamaica, highlights a dense chain of informal and formal linkages. The industry was established in 1956 coinciding with widespread migration to the UK and the creation of demand for canned and preserved versions of ackee for export. There is a long chain of production that includes both informal and informal linkages, as well as legal and illegal processes.

Within Jamaica, there are four main levels in the production chain reflecting various levels of informality and illegality. At the bottom of the chain is a loose collection of ackee pickers who supply truckers and traders. Because ackee trees grow wild rather than in orchard settings, the fruit is picked by casual labourers. Sometimes, truckers and traders make door-to-door calls asking for the fruit, while in other cases they assemble groups of male teenagers with a knowledge of where the best ackee trees are located; these young men will guide the truckers to the trees where they then raid them (illegally if the land belongs to someone else). The truckers and traders supply the ackees to small, regulated firms which check, sort, clean and can the ackees in their small processing factories. Some of these small firms use their own labels and product names, while others label the cans in the names of the larger distributors. These distributors are the last level in the chain of production, and these firms are responsible for distributing the cans within the country and for export to the UK, Canada and the US. Although these distributors are formal in terms of their regulation and scale, export can be illegal when it involves the US – the largest export market. This is because the fruit contains hypoglicin – a chemical substance that is banned by the US Food and Drug Administration. Therefore, an illegal system of international distribution has been established among formal and informal firms in the US predicated on networks of trust and distrust of importers in the US.

Source: Gordon *et al.* (1997).

---

recent increases in informal employment in Middle America and elsewhere, and in turn is linked with recession and restructuring. Bearing in mind the difficulties in collecting accurate employment figures, especially among women workers whose economic contributions are notoriously under-reported (Chant 1999b), these relationships can be seen by continued increases in female labour force participation during the so-called 'lost decade' of the 1980s until the present day (Chant, this volume).

In a range of Middle American countries, with the exception of Haiti, women's share of the labour force increased between 1980 and 1999, being particularly marked in El Salvador and Nicaragua (Table 5.3). While some women have secured jobs in the formal export-manufacturing sector of

Table 5.3 Women's share of the labour force in 1980 and 1999 in selected Middle American countries

Country	*Female percentage of labour force*	
	1980	*1999*
Dominican Republic	25	30
El Salvador	27	36
Guatemala	22	28
Haiti	45	43
Honduras	25	31
Jamaica	46	46
Mexico	27	33
Nicaragua	28	35
Panama	30	35

Source: World Bank (2000: 278–279).

Middle American economies, notably in El Salvador, the Dominican Republic, Mexico and Guatemala, the actual numbers involved have been small (Safa 1995). Greater numbers of women have become involved in informal subcontracted piecework that forms the classic link between the formal and informal sectors. Indeed, it is flexible labour that accounts for the upward trend in the female share of the labour force in Middle America and elsewhere (Standing 1999: 588). Thus, the bulk of this expanding flexible labour is in the informal sector, with recent entrants comprising mostly female workers. While it is unclear why there has been a small decline in Haiti, it may be related to the lack of export-manufacturing activity that provides significant direct and indirect employment opportunities for women.

While there are obviously context-specific variations, women informal sector workers tend to be concentrated in activities that require the least skills and experience, the least start-up costs and access to capital, and in jobs that allow them to combine reproductive activities, such as house-work and childcare, with income-generating activities (Scott 1995). These patterns can be partly explained because women invariably have to work out of economic necessity, especially since the 1980s (see for example Chant 1996; González de la Rocha 1988 on Mexico). With making a living the primary motivating factor, women thus develop whatever activities or strategies they can. Men on the other hand are most likely to be involved in more highly remunerated occupations, involving more capitalised goods with much higher returns. They are also often more likely to work at a distance from their home (Chant 1999b).

While these refer to horizontal divisions, there are also vertical gender divisions within small-scale businesses. For instance, in family businesses, men are often in charge of decision-making and finances, whereas women are more likely to be unpaid workers (see Lloyd Evans 1998 on Trinidad; also Box 5.2). In the light of this, it is not surprising that women tend to

Plate 5.2 Informal sector lottery seller, San José, Costa Rica
Photo: Cathy McIlwaine

have lower earnings than men in the informal sector. In the Dominican Republic, 62 per cent of female informal workers earn below the poverty level, compared with only 35 per cent of male informal workers (Lozano 1997: 163).

It should also be reiterated that women tend to be concentrated in domestic service activities. Bearing in mind problems with data reliability, it has been estimated that in 1993, 25 per cent of women wage-workers in Honduras were domestic workers (López de Mazier 1997: 237). Similarly, in El Salvador, also in 1993, 14.4 per cent of all female workers were domestic servants (Gutiérrez Castillo 1997: 149). Also important is that domestic service is highly diverse with respect to informality. In some instances, workers may receive a monthly wage and various statutory benefits, while others may work on an hourly basis in a range of different jobs. Overall, however, domestic service is usually highly exploitative of women who work long hours, with low pay and severe restrictions on their mobility if they are 'live-ins'.

Along with gender, other axes of differentiation are also important. In a classic study of Mexico City, Lourdes Arizpe (1977) examines the differences between middle- and working-class women making a living in the informal sector. As would be expected, she notes that middle-class women tend to be engaged in more 'up-market' activities such as tutoring, whereas working-class women work mainly in domestic service as well as street trading of edible items such as sweets and chewing gum. Arizpe also

Box 5.2 Gender, ethnicity and the informal sector in Trinidad

An in-depth national level study of petty commodity trading undertaken in the early 1990s in Trinidad highlighted important and pervasive occupational divisions on grounds of gender and ethnicity. In a survey of 740 food traders, two-thirds of the sample were male; a pattern unusual in the Caribbean as a whole. Among these traders, there were distinct divisions of occupations with women being involved in food trading or higglering, and men mainly working as highway traders. Higglers, who sell both domestic and imported food, tended to operate from the capital, Port of Spain's, central market as well as from regional markets. Among female informal workers, female higglering was a career, compared with other female food traders who were concentrated in low status, irregular occupations such as casual selling of small amounts of local products such as limes and mangoes from makeshift street stalls. Confectionary trading (nuts and cooked food) was also female-dominated and low status, mainly because it involves perishable goods with few capital inputs. Male food traders, on the other hand, worked mainly in family-run, fairly profitable businesses. Moreover, many of these businesses involved men travelling to sell the goods, whereas women remained in the home even though they contributed substantially to the success of the business. In a further sample of 170 urban workers engaged in the sale of non-food items, similar gender divisions of labour prevailed. Female traders were predominantly employed in small businesses which utilised reproductive skills such as dress-making and clothing and cosmetic trading. In contrast, men ran more successful businesses in repair, electronics, imports and transport.

There were also marked ethnic patterns in informal activities. Although Indo-Trinidadians (people of Asian origin) are just under half the national population, they are the majority of highway food traders (70 per cent). This group were primarily involved in the family-run businesses where women received little recognition for their contributions. Indeed, Indo-Trinidadian women were conspicuous in their absence from the public sphere of trading in the country as whole. In contrast, the vast majority of the female higglers were Afro-Trinidadian (of African origin). Among the non-food traders, based mainly in the urban areas, 85 per cent were Afro-Trinidadian rather than Indo-Trinidadian, with roughly equal numbers of men and women. Afro-Trinidadian women had considerably more freedom in their informal sector activities than their Indo-Trinidadian counterparts. Indeed, some of the former were engaged in 'suitcase trading' involving travel to the US, Puerto Rico and other Caribbean islands to purchase consumer goods such as clothes, music and so on, for sale back home.

Sources: Lloyd Evans (1997, 1998).

highlights age differentials; young single women are more likely to work as domestic servants, whereas older and elderly women are concentrated in petty trading and market work.

Another study of piecework from rural Honduras discusses the importance of gender, age and lifecycle in informal employment. Pérez Sáinz (1998) examines a Honduran–US manufacturing company that employs women under subcontracting arrangements to sew baseballs. The company deliberately decentralised after a trade union was established, and explicitly employs mature women at the peak of their reproductive years in the knowledge that, because they have difficulty entering the labour market, they will be grateful for any work they can get (Pérez Sáinz 1998).

Ethnicity is also significant, although there are comparatively few studies on the relationships of this with the informal sector in Middle America. Those that do examine how different ethnic groups are located within the labour market usually highlight how ethnic minority groups tend to be concentrated in the most precarious activities and invariably in the informal sector. For instance, in Guatemala, Pérez Sáinz (1997b) notes that indigenous groups are most likely to work in the informal sector in low wage and low skill activities, with women and children particularly concentrated in this sphere. Lloyd Evans' (1998) study of Trinidad also disaggregates various types of informal work along lines of ethnicity (and gender). She shows how different ethnic groups (Indo-Trinidadians and Afro-Trinidadians) are concentrated in distinct activities within the informal sector (see Box 5.2).

An increasing number of studies of the informal sector in Middle America have discussed – explicitly and implicitly – the issue of social capital (Lloyd Evans 1997; Portes and Itzigsohn 1997; Portes and Landolt 2000). These have examined the ways in which particular groups utilise various forms of social capital in ensuring that their businesses function, which, in the context of the informal sector, refers to networks of reciprocity or sharing of knowledge, capital loans and credit or business contacts (see McIlwaine, this volume, for more complete definitions). For example, among the Mayan clothing manufacturers in Guatemala mentioned earlier, micro-entrepreneurs are bound together by their ethnicity, assisting each other in various ways. Apprenticeships, for example, take place in workshops of family or friends, after which capital or credit is obtained from kin. Once the workshop is up and running, friends and family introduce the new workers to the international firms involved in the process of manufacture (Pérez Sáinz 1997a).

Children and the informal sector

As well as gender and ethnic variations within the informal sector, there are differences along lines of age. Especially important is the issue of child labour as children usually work in the most exploitative echelons of informal activities. Although some children may work as operatives in factories or *maquilas* (or more commonly as subcontractors to these factories), the majority are engaged in less formal occupations such as urban street selling. Children's experiences in the informal sector can be conceptualised mainly as 'child labour' which is waged work, although in some instances it may involve 'child work' referring to unwaged work usually carried out in the home.

While child labour is not as prevalent in Latin and Middle America as in countries of Africa and Asia, and data are notoriously inaccurate, estimates suggest that 20 million children in Latin America are working (Osorio Ponce 1998: 66). Furthermore, among children aged 10 to 14 years of age, it is estimated that 9 per cent were working in 1997, representing a decline from 1980 when 13 per cent were working (World Bank 1999: 194–5). However, in some countries in the region, child labour is considerably more widespread, especially where economic necessity is greatest. In Haiti, for example, despite a decline from around one-third in 1980, as many as 24 per cent of all children were working in 1997 (ibid.). Even in Costa Rica, which is not noted for its high levels of child labour, a recent interagency study suggests that in the canton of Upala on the Nicaraguan border, 23 per cent of young persons between the ages of 14 and 17 years are working, and four-fifths of them do not study (Osorio Ponce 1998: 66). It is also argued that neoliberalism and unequal trade resulting from economic globalisation has exacerbated child labour and child work even further (Robertson 1994). With reference to Latin America as a whole, Green (1998: 35) argues: 'In this brave new world of "flexible working patterns", children are often perfect employees – the cheapest to hire, the easiest to fire and the least likely to protest.'

While poverty and economic crisis have led to an increased need for children to earn a living as family resources are stretched to their limits, this situation has also generated an increase in street children (Bernat 1999 on Haiti). The reasons for children ending up on the streets are manifold, but are most commonly associated with abuse, loss of parents, or children being thrown out by families who can no longer feed them. Moreover, some children also display considerable agency in instigating their departure from adult control (see Jones 1997 on Mexico).

While many street children engage in 'above board' informal livelihood strategies, they are often associated in the public eye with petty crime and violence, although in reality they are both victims and perpetrators. Because of this association with crime, they are often targeted by authorities and private individuals. For example, in Guatemala City, street children have been targets of 'social cleansing' (killing) because of their assumed involvement in petty crime, prostitution, and begging, with the police being identified as the main perpetrators (Human Rights Watch, 1997; Moser and McIlwaine 2001). Street children in Port-au-Prince, Haiti, face similar dangers, being hounded and criminalized by state authorities (Bernat 1999). Increasingly, street children and child workers are the focus of international policymaking, especially on the part of global institutions such as the ILO and UNICEF. However, a paternalistic stance is often taken to child workers, regarding them as passive victims of an unfair global system in need of protection (Jones 1997), rather than as social actors on their own terms (Ennew 1994). UNICEF has suggested differentiating interventions for child labour by age group. For those under 12 years, all

Plate 5.3   Informal artisans (electricians, plumbers, etc.) touting their skills, the zócolo, Mexico City
Photo: Cathy McIlwaine

work should be abolished and the minimum age raised to 15. Meanwhile, child workers aged 13 and 14 should receive professional training and apprenticeships. Young workers aged 15 to 17 years should be covered by legal protection and also receive professional assistance (Osorio Ponce 1998: 76).

Policy perspectives on the informal sector

Just as there are policy debates revolving around the most suitable interventions to address child labour, so there is a range of approaches dealing with the informal sector. Although this chapter has considered both formal and informal employment, the fact that the formal sector has become increasingly informalised, and because informal employment has grown substantially and involves such as large proportion of the labour force of the region, the focus here is on informal sector policy debates.

Approaches towards the informal sector have been formulated only relatively recently in Latin and Middle America for two main reasons. First, it was initially believed that labour surpluses would be absorbed into the formal sector, a process that did occur in some Latin American countries between 1950 and 1980 (Tokman 1989). Second, and perhaps more pervasive, was the notion that the informal sector was 'parasitic' and 'unproductive'. Therefore, few policymakers were willing to promote a sector that was viewed in such a negative manner (Bromley 1997).

Consequently, informal sector workers and entrepreneurs have for a long time faced serious barriers to survival and growth. One major barrier is the disproportionate attention that governments have given to promoting the large-scale capital-intensive sector. Policies favouring the formal sector (at the expense of the informal sector) have invariably involved credit transfers, direct and indirect market protection such as tariffs, quotas and so on (Chickering and Salahdine 1991: 186). The second major barrier to informal sector growth is the way in which excessive regulation and bureaucracy made it too time-consuming and costly to become legal (Tokman 1991; see also Bromley 1997). Beyond this, other obstacles to informal workers have included harassment or victimisation (Thomas 1996: 56–7).

More in-depth consideration of the nature of these obstacles forms the basis of a recent theoretical and policy debate that has resonated throughout the developing world in the last two decades. This revolves around the notion of whether the informal sector is a poverty trap or a sphere of latent entrepreneurial economic dynamism. This debate draws partly on Hart's early work that sought to overturn many of the negative notions about the informal sector. It also draws on many studies that show that informal workers often earn more than salaried workers, especially since the decline of real wages in the formal sector (see Gilbert and Gugler 1992; Thomas 1995). Furthermore, it is often argued that self-employment can be a source of pride or prestige, as well as providing flexibility to changing economic demands and family circumstances, especially as people learn skills over time (often in the formal sector). However, the key difference between Hart's analysis and the more recent debates about 'accumulation' versus 'survival' is that the sector is not seen as originating from surplus labour supply, but from excessive regulations in the economy (Portes and Schauffler 1993: 39–40).

No discussion of this debate is complete without reference to Hernando de Soto's book *The Other Path* (1989). De Soto, a Peruvian economist and member of the Peruvian Institute for Liberty and Democracy, hails the informal sector as the potential answer to the employment problems in the developing world. As noted above, his focus is on what he perceives as the injustice of excessive regulation created by governments in favour of society's powerful and dominant groups (Bromley 1997). In arguing that legal systems function in the interests of the economic elite, de Soto suggests that illegality is the only viable alternative for the poor. However, this 'illegality' is conceived more as 'non-conformity' with rules and regulations

127

rather than any criminal act. Indeed, de Soto argues that informal sector work provides an important alternative to crime, as well as relieving unemployment. Furthermore, his key idea is that governments should encourage and protect informal entrepreneurs, so that their entrepreneurial talent can flourish (de Soto 1989).

While de Soto's work is often criticised for being a restatement of 1970s research that argued that the informal sector was a 'panacea' for employment problems, his theoretical insights and policy recommendations have appealed to both left and right of the political spectrum throughout the developing world in the 1990s (Portes and Itzigsohn 1997). However, neoliberals have perhaps been the most enthusiastic about his ideas of deregulation, as they concur closely with their propositions incorporated within SAPs promoted by the IMF and World Bank. Indeed, although de Soto's positive attitude towards the informal sector is generally welcomed, the desirability of further liberalisation of markets and production processes is in greater question. De Soto's arguments, for example, can be used to provide a rationale for condoning deteriorating work conditions, greater incidence of worker abuse, minimal wages, and disincentives for employers to provide training and/or engage in technological innovation (Portes and Schauffler 1993). Other flaws in his theory include the fact that the idea that an excessive bureaucracy generates an informal sector is not borne out in wealthier economies of the world (ibid.). Greater tolerance of poor working conditions in the informal sector can also be politically expedient as it helps to depress unemployment figures.

The aspect of de Soto's theory that has perhaps received most attention is the assumption that informal sector workers are potential entrepreneurs. This is hard to uphold given the empirical evidence from Middle America and beyond that shows that the majority of informal sector workers can merely survive in the sector rather than flourish (Thomas 1995). Even if it is recognised that the potential for making a profit depends on the activity, and often the extent to which small-scale enterprises are linked with larger firms, in an example from a low-income area in Guatemala City, Bastos and Camus (cited in Pérez Sáinz 1997b: 87) found that only 7.3 per cent of all informal businesses could be viewed as 'dynamic' or driven by 'entrepreneurial rationality'. They also found that the poorer the area, the greater the likelihood that informal enterprises would be based on survival alone. The only dynamic activities were those that diversified beyond commerce, were located outside the home, employed paid workers, and employed fewer women or family members.

Returning to specific policies, it is generally agreed that interventions to assist the informal sector are required, even though it is recognised that fully protected employment should really be the ultimate goal. With reference to the Caribbean, Portes and Itzigsohn (1997: 241–3) draw on some of de Soto's basic ideas, identifying the importance of removing constraints to informal sector expansion. These obstacles include: first, lack of capital due to limited access to mainstream financial institutions; second, concentration in highly competitive low-income markets with few possibilities for growth; third, excessive use of restrictive agencies or middlemen in some sectors of the informal economy that block access to more dynamic sectors of the economy; fourth, social atomisation of informal entrepreneurs due to the irregular and/or chaotic nature of supplies; finally, traditional ethics, such as a 'craftsman ethic' that prevents some informal entrepreneurs from changing their traditional methods of production.

On the supply side of the labour market, policies have increasingly focused on education and training to promote the diversification of the informal sector, as well as enhanced access to credit, usually in the guise of micro-enterprise development. This often entails assistance in management, marketing and packaging, and measures to promote greater health and safety. These initiatives are particularly relevant for groups within the informal sector such as ambulant traders and food vendors, where women are often a large percentage of operatives (Blumberg 1995; Tinker 1997). Finally, there have also been suggestions for reorientating policies away from individual firms or workers as a means of utilising the social networks and social capital that frequently fuel the operation of the informal sector (Portes and Itzigsohn 1997; Portes and Landolt 2000).

Conclusion

The informal sector is likely to form a major aspect of urban economies in Middle America for decades to come. The linkages between the formal and informal sectors are also likely to multiply if deregulation continues apace. With increasing economic diversification in the region, especially towards export-manufacturing, many forms of subcontracting and piece-work are likely to open up for low-income groups. However, while these aspects of the globalisation process may provide some opportunities, it should also be remembered that pieceworkers, in particular, are working at the disadvantaged end of a much wider chain of exploitative links where benefits accrue primarily to transnational corporations. Besides sub-contracted labour, other forms of informal employment are also likely to continue to grow in the future. Some of this labour will involve establishment of successful small-scale businesses. However, much informal work will provide only a rudimentary way of making a living for many people in the region. As noted throughout the chapter, there are important gender, ethnic and age variations in terms of the types of activities in which people engage. Bearing in mind huge diversity in these activities, it has been repeatedly shown that women, children and ethnic minority groups are the most likely to be entrenched in subsistence activities making only an exiguous living in whatever way possible. A social justice approach to employment and livelihoods would clearly attempt to rectify the imbalance and concentrate policy resources on converting these subsistence activities into more rewarding entrepreneurial concerns. If market-driven economic tendencies continue, however, the likelihood is that further polarisation will occur, and the informal sector's position as a poverty trap will deepen.

Further reading

Moser, C. (1978) Informal sector or petty commodity production? Dualism or dependence in urban development, *World Development* **6**: 135–178.
A classic paper outlining the conceptual linkages between the formal and informal sectors of the labour market.

Portés, A., Dore-Cabral, C. and **Landolt, P.** (eds) (1997) *The Urban Caribbean: Transition to a New Global Economy*. The Johns Hopkins University Press, Baltimore.
An interesting collection with an underlying theme of whether the informal sector is an arena of growth or survival. Case studies from Costa Rica, Haiti, Guatemala, the Dominican Republic and Jamaica.

Portes, A. and **Schauffler, R.** (1993) Competing perspectives on the Latin American informal sector, *Population and Development Review* **19**(3): 33–60.
A useful outline of some of the key definitional debates and approaches on the informal sector from a Latin American viewpoint.

Thomas, J.J. (1995) *Surviving in the City: the Urban Informal Sector in Latin America*. Pluto Press, London.
An excellent summary of all the major debates on the informal sector. The book includes conceptual and empirical material presented in an accessible manner.

Standing, G. and **Tokman, V.** (eds) (1991) *Towards Social Adjustment: Labour Market Issues in Structural Adjustment*. International Labour Office, Geneva.
Includes a series of chapters on the informal sector, examining how the sector has changed in the light of structural adjustment policies.

References

Arizpe, L. (1977) Women in the informal labor sector: the case of Mexico City, *Signs* **3**(1), reprinted in Visvanathan, N., Duggan, L., Nisonoff, L. and Wiegersma, N. (1997) (eds) *The Women, Gender and Development Reader*. Zed, London: 230–237.

Baden, S. (1993) *The Impact of Recession and Structural Adjustment on Women's Work in Developing Countries*. Bridge Report No.2, Institute of Development Studies, Brighton, Sussex.

Benería, L. and **Roldán M.** (1987) *The Crossroads of Class and Gender: Industrial Homework, Subcontracting and Household Dynamics in Mexico City*. University of Chicago Press, Chicago.

Bernat, J.C. (1999) Children and the politics of violence in Haitian context, *Critique of Anthropology* **19**(2): 121–138.

Blumberg, R.L. (1995) Gender, microenterprise, performance and power: case studies from the Dominican Republic, Ecuador, Guatemala and

Swaziland, in C. Bose and E. Acosta-Belén, E. (eds), *Women in the Latin American Development Process*. Temple University Press, Philadelphia: 194–226.

Bromley, R. (1979) (ed.) *The Urban Informal Sector: Critical Perspectives on Employment and Housing Policies*. Pergamon Press, Oxford.

—— (1997) Working in the streets of Cali, Colombia: survival strategy, necessity or unavoidable evil?, in Gugler, J. (ed.) *Cities in the Developing World: Issues, Theory and Policy*. Oxford University Press, Oxford: 124–138.

Bulmer-Thomas, V. (1996) Conclusions, in Bulmer-Thomas, V. (ed.) *The New Economic Model in Latin America and its Impact in Income Distribution and Poverty*. Macmillan, Basingstoke: 296–327.

Chant, S. (1991) *Women and Survival in Mexican Cities: Perspectives on Gender, Labour Markets and Low-income Households*. Manchester University Press, Manchester.

—— (1996) Women's roles in recession and economic restructuring in Mexico and the Philippines, *Geoforum* **27**(3): 297–327.

—— (1999a) Population, migration, employment and gender, in Gwynne, R. and Kay, C. (eds) *Latin America Transformed: Globalisation and Modernity*. Edward Arnold, London: 226–269.

—— (1999b) Informal sector activity in the Third World City, in Pacione, M. (ed.) *Applied Geography: an Introduction to Useful Research in Physical, Environmental and Human Geography*. Routledge, London: 509–527.

Chickering, A.L. and **Salahdine, M.** (1991) The informal sector's search for self-governance, in Chickering, A.L. and Salahdine, M. (eds) *The Silent Revolution: the Informal Sector in Five Asian and Near Eastern Countries*. International Center for Economic Growth, San Francisco: 185–211.

Cubitt, T. (1995) *Latin American Society* (2nd edn.). Longman, London.

Ennew, J. (1994) *Street Children and Working Children: a Guide to Planning*. Save the Children, London.

Escobar Latapí, A. (1988) The rise and fall of an urban labour market: economic crisis and the fate of small workshops in Guadalajara, Mexico, *Bulletin of Latin American Research* **7**(2): 183–205.

Escobar Latapí, A. and **González de la Rocha, M.** (1995) Crisis, restructuring and urban poverty in Mexico, *Environment and Urbanisation* **7**(1): 57–76.

Gilbert, A. (1994) Third World cities: poverty, employment, gender roles and the environment during a time of restructuring, *Urban Studies* **30**: 721–740.

Gilbert, A. and **Gugler, J.** (1992) *Cities, Poverty and Development: Urbanisation in the Third World*. Oxford University Press, Oxford.

González de la Rocha, M. (1988) Economic crisis, domestic reorganisation and women's work in Guadalajara, Mexico, *Bulletin of Latin American Research* **7**(2): 207–223.

Gordon, D., Anderson, P. and **Robotham, D.** (1997) Jamaica: urbanization during the years of the crisis, in Portés, A., Dore-Cabral, C. and Landolt, P. (eds) *The Urban Caribbean: Transition to a New Global Economy.* The Johns Hopkins University Press, Baltimore: 190–223.

Green, D. (1995) *Silent Revolution: the Rise of Market Economics in Latin America.* Cassell in asociation with Latin America Bureau, London.

—— (1996) Latin America: neoliberal failure and the search for alternatives, *Third World Quarterly.* **17**(1): 109–122.

—— (1998) *Hidden Lives: Voices of Children in Latin America and the Caribbean.* Routledge, London.

Gutiérrez Castilo, M. (1997) Aspectos de género de la economía de El Salvador, in Elson, D., Fauné, M.A., Gideon, J., Gutiérrez, M., López de Mazier, Armida and Sacayón, E. (eds) *Crecer con la Mujer: Oportunidades para el Desarrollo Económico Centroamericano.* Embajada Real de los Países Bajos, San José: 127–171.

Hart, K. (1973) Informal income opportunities and urban employment in Ghana, in Jolly, R., de Kadt, E., Singer. H. and Wilson, F. (eds) *Third World Employment.* Penguin, Harmondsworth: 66–70.

Human Rights Watch/Americas (1997) *Guatemala's Forgotten Children: Police Violence and Abuses in Detention.* Human Rights Watch, New York.

Itzigsohn, J. (1997) The Dominican Republic: politic-economic transformation, employment, and poverty, in Tardanico, R. and Menjívar Larín, R. (eds) *Global Restructuring, Employment, and Social Inequality in Urban Latin America.* North–South Center, University of Miami, Coral Gables, Florida: 47–72.

Jones, G.A. (1997) Junto con los niños: street children in Mexico, *Development in Practice* **1**(1): 39–49.

Lloyd Evans, S. (1997) Gender, ethnicity and social capital in Trinidad's informal sector. Paper presented at the Annual Conference of the Society for Latin American Studies, 4–6 April 1997, St. Andrew's University.

—— (1998) Gender, ethnicity and small business development in Trinidad: prospects for sustainable job creation, in McGregor, D., Barker, D. and Lloyd Evans, S. (eds) *Sustainability and Development in the Caribbean: Geographical Perspectives.* University of West Indies Press, Mona, Jamaica: 3–25.

López de Mazier, A. (1997) La mujer, principal sostén del modelo económico de Honduras: un análisis de género de la economía Hondurena, in Elson, D., Fauné, M.A., Gideon, J., Gutiérrez, M., López de Mazier, Armida and Sacayón, E. (eds) *Crecer con la Mujer: Oportunidades para el Desarrollo Económico Centroamericano.* Embajada Real de los Países Bajos, San José: 215–252.

Lozano, W. (1997) Dominican Republic: informal economy, the state and the urban poor, in Portés, A., Dore-Cabral, C. and Landolt, P. (eds) *The Urban Caribbean: Transition to a New Global Economy*. The Johns Hopkins University Press, Baltimore, 153–189.

Moghadam, V. (1995) Gender aspects of employment and unemployment in global perspective, in Simai, M. with Moghadam, V. and Kuddo, A. (eds) *Global Employment: an International Investigation into the Future of Work*. Zed, in association with United Nations University, World Institute for Development Economics Research, London: 111–139.

Moser, C. (1978) Informal sector or petty commodity production? Dualism or dependence in urban development, *World Development* 6: 135–178.

—— (1998) The asset vulnerability framework: reassessing urban poverty reduction strategies, *World Development* **26**(1): 1–9.

Moser, C. and **McIlwaine, C.** (2001) *Violence in a Post-Conflict Context: Urban Poor Perceptions from Guatemala*. World Bank, Washington, DC.

Osorio Ponce, R. (1998) Child-adolescent labour and the rights of boys, girls and adolescents, in UNICEF (ed.) *Adolescence, Child Rights and Urban Poverty in Costa Rica*. UNICEF/HABITAT, San José: 65–80.

Pérez Sáinz, J.P. (1997a) Crisis, restructuring, and employment in Guatemala, in Tardanico, R. and Menjívar Larín, R. (eds) *Global Restructuring, Employment, and Social Inequality in Urban Latin America*. North–South Center, University of Miami, Coral Gables, Florida: 73–94.

—— (1997b) Guatemala: two faces of the metropolitan area, in Tardanico, R. and Menjívar Larín, R. (eds) *Global Restructuring, Employment, and Social Inequality in Urban Latin America*. North–South Center, University of Miami, Coral Gables, Florida: 124–152.

—— (1998) The new faces of informality in Central America, *Journal of Latin American Studies* **30**: 157–179.

Portes, A. and **Itzigsohn, J.** (1997) Coping with change: the politics and economics of urban poverty, in Tardanico, R. and Menjívar Larín, R. (eds) *Global Restructuring, Employment, and Social Inequality in Urban Latin America*. North–South Center, University of Miami, Coral Gables, Florida: 227–248.

Portes, A. and **Landolt, P.** (2000) Social capital: promise and pitfalls of its role in development, *Journal of Latin American Studies* **32**: 529–547.

Portes, A. and **Schauffler, R.** (1993) Competing perspectives on the Latin American informal sector, *Population and Development Review* **19**(3): 33–60.

Potter, R. and **Lloyd Evans, S.** (1998) *The City in the Developing World*. Addison Wesley Longman, Harlow.

Roberts, B. (1994) Informal economy and family strategies, *International Journal of Urban and Regional Research* **18**(1): 6–23.

—— (1995) *The Making of Citizens: Cities of Peasants Revisited.* Edward Arnold, London.

Robertson, A. (1994) Free trade or fair trade, *Anti-Slavery International Reporter* **13**(9): 63–64.

Safa, H. (1995) Economic restructuring and gender subordination, *Latin American Perspectives* **22**(2): 32–50.

Scott, A. MacEwen (1995) Informal sector or female sector? Gender bias in urban labour market model, in Elson, D. (ed.) *Male Bias in the Development Process* (2nd edn.). Manchester University Press, Manchester: 105–132.

de Soto H. (1989) *The Other Path: the Invisible Revolution in the Third World.* Harper and Row, New York.

Standing, G. (1989) Global feminisation through flexible labour, *World Development* **17**(7): 1077–95.

—— (1999) Global feminisation through flexible labour: a theme revisited. *World Development* **27**(3): 583–602.

Tardanico, R. (1997) From crisis to restructuring: Latin American transformations and urban employment in world perspective, in Tardanico, and Menjívar Larín, R. (eds) *Global Restructuring, Employment, and Social Inequality in Urban Latin America.* North–South Center, University of Miami, Coral Gables, Florida: 1–46.

Thomas, J.J. (1995) *Surviving in the City: the Urban Informal Sector in Latin America.* Pluto Press, London.

—— (1996) The new economic model and labour markets in Latin America, in Bulmer-Thomas, V. (ed.) *The New Economic Model in Latin America and its Impact on Income Distribution and Poverty.* Macmillan, in association with the Institute of Latin American Studies, University of London, Basingtoke: 79–102.

—— (1999) El mercado laboral y el empleo, in Crabtree, J. and Thomas, J. (eds) *El Perú de Fujimori.* Universidad del Pacífico, Lima: 255–296.

Tinker, I. (1997) *Street Foods: Urban Food and Employment in Developing Countries.* Oxford University Press, New York and Oxford.

Tokman, V. (1989) Policies for a heterogeneous informal sector in Latin America, *World Development* **17**(7): 1067–1076.

—— (1991) The informal sector in Latin America: from underground to legality, in Standing, G. and Tokman, V. (eds) *Towards Social Adjustment: Labour Market Issues in Structural Adjustment.* International Labour Office, Geneva: 141–157.

United Nations (UN) (1995) *The World's Women 1995: Trends and Statistics.* UN, New York.

United Nations Development Programme (UNDP) (2000) *Human Development Report 2000.* Oxford University Press, New York.

United Nations Economic Commission for Latin America and the Caribbean (ECLAC) (1994) *Social Panorama of Latin America.* ECLAC, Santiago.

Ward, K. and **Pyle, J.** (1995) Gender, industrialisation, transnational corporations and development: an overview of trends and patterns, in Bose, C. and Acosta-Belén, E. (eds) *Women in the Latin American Development Process.* Temple University Press, Philadelphia: 37–64.

Wield, D. and **Chataway, J.** (2000) Unemployment and making a living, in Allen, T. and Thomas, A. (eds) *Poverty and Development into the 21st Century.* Oxford University Press, Oxford: 99–124.

World Bank (1999) *World Development Report 1999.* Oxford University Press, Oxford and New York.

—— (2000) *World Development Report 2000.* Oxford University Press, Oxford and New York.

CHAPTER 6

Open for business: strategies for economic diversification

Katie Willis

Introduction

While globalisation is argued to be a recent phenomenon (see Willis and McIlwaine, this volume), it is clear that Middle America has been embedded in the world economy for many centuries. However, for much of this period such integration has been related to agricultural and mining products, rather than manufactured goods or service provision. This chapter examines the rise of industrial production in the region, how this was linked to national development strategies, and how external factors have influenced the nature of industrialisation from a broadly import-substitution model to an export-led manufacturing sector, involving large amounts of international capital. This move towards a more open economy is also found within the service sector, two important elements of which are tourism and financial services. These will be examined later in the chapter. A key aim of this chapter is to demonstrate the heterogeneity of industrial and service development due to local factors, and also local responses to globalisation within the broad trends of economic diversification.

Manufacturing in the region

Before embarking on a description of recent industrialisation policies in the region, it is useful to have an overview of the patterns of manufacturing and industrial activity (see Table 6.1). With the exception of the Dominican Republic, Trinidad and Tobago, and Jamaica (where the contribution of industry to GDP was higher than for other countries in the region), industry represented 20 to 23 per cent of GDP in 1997, demonstrating its important contribution to the region's economies. However, the 'industry' category includes a range of activities, including processing of agricultural and mineral products, so the manufacturing column gives a better idea of the way in which greater 'value-added' production, such as assembly factories, fits into the region's economic profile. Industry can provide an important source of employment, although the impact varies depending on the degree of capital-intensiveness. Where industrial development has been unable to incorporate the ranks of the unemployed, workers have had to devise their own livelihood strategies, including working in the informal sector (see McIlwaine *et al.*, this volume). Manufacturing is also a key source of foreign currency, with manufactures representing over 50 per cent of merchandise exports in Barbados, Jamaica and Mexico, and making significant contributions to the export profile in other countries (see Table 6.1).

The three exceptions mentioned above represent countries where primary processing plays a crucial role in the economy; in Jamaica through the bauxite industry, the Dominican Republic's sugar-processing industry and Trinidad and Tobago's petroleum industry. However, all three

Table 6.1 Manufacturing indicators

Country	Industry value added as % of GDP (1997)[a]	Manufacturing value added as % of GDP (1998)[b]	% labour force in industry (1990)[c]	Manufactures as % of merchandise exports (1997)[d]
Bahamas	n.a.	n.a.	15	n.a.
Barbados	n.a.	n.a.	23	54
Belize	28	n.a.	19	13
Costa Rica	23	16	27	**25**
Cuba	n.a.	n.a.	30	n.a.
Dominica	21	n.a.	n.a.	**49**
Dominican Republic	32	17	29	n.a.
El Salvador	28	22	21	39
Grenada	20	n.a.	n.a.	**13**
Guatemala	20	13	17	30
Guyana	n.a.	n.a.	25	n.a.
Haiti	20	n.a.	9	n.a.
Honduras	28	18	20	27
Jamaica	35	16	23	**69**
Mexico	26	20	24	81
Nicaragua	22	16	26	25
St Kitts & Nevis	25	n.a.	n.a.	34
St Lucia	20	n.a.	n.a.	**25**
St Vincent and Grenadines	25	n.a.	n.a.	13
Suriname	n.a.	n.a.	18	n.a.
Trinidad & Tobago	46	n.a.	n.a.	44

Sources: [a] UNDP (1999: Table 12, 184–186); [b] World Bank (2000: Table 12, 252–253); [c] UNDP (1998: Table 16, 164–165); [d] UNDP (1999: Table A1.1, 45–48).
Notes
n.a. = not available
Figures in bold refer to 1996 data

countries have also used this primary processing as a base for the development of other forms of factory production in sectors such as textiles and electronics.

Going it alone?: Import substitution industrialisation

While the majority of this chapter will be geared towards recent trends associated with manufacturing for a global export market, this approach to national development needs to be considered in relation to previous industrialisation attempts made by a number of the region's governments. This strategy of import substitution industrialisation (ISI) was associated with the ideas of the dependency school linked with the work of the UN's Economic Commission for Latin America (ECLA) (see Willis and McIlwaine, this volume). In an attempt to restructure the nature of economic relations with core economies, some governments sought to promote industrial development for the domestic market. This would reduce dependence on imported goods, so helping the balance of payments, and would also provide employment. While other sectors (especially agriculture, but also tourism) were also given attention, many governments viewed manufacturing as the key to 'development'. However, it must be stressed that although production was to be geared towards the domestic market, foreign investment was encouraged (Bulmer-Thomas 1987).

From the 1930s onwards, national governments adopted a range of policies to encourage domestic industrialisation. These measures included increasing tariffs on imported goods. In Central America this was largely to help raise revenue, rather than promote import substitution (Bulmer-Thomas 1987), but in Mexico the ISI strategy was more explicit and government-led. Behind the protective walls of import tariff barriers, some industrial development took place. In Mexico, industry's share of total output rose from 21.5 per cent in 1950 to 24 per cent in 1960, and 24.9 per cent in 1970 (Lustig 1992).

However, it was clear that this success was rather fragmented, both spatially and temporally. Some of the region's economies were in a better position to benefit from ISI, largely because of the size of the domestic market and the development of infrastructure. The importance of infrastructure in the expansion of production can also be seen in relation to agriculture (see Thorpe and Bennett, this volume). Regional co-operation, such as the Central American Common Market (CACM) established in 1960, was proposed as a way of escaping the limited size of domestic markets. Overall, these attempts at co-operation often faltered, largely because countries were producing similar products (see below). In addition, existing social structures, particularly the power of the agricultural elite, hindered the promotion of industry, as such groups were unwilling to relinquish the positions of economic dominance (Bulmer-Thomas 1987). Finally, although some consumer goods were relatively straightforward to produce, attempts to replace the import of other goods (particularly industrial machinery) highlighted the limits of ISI at that time.

Export-oriented industrialisation

While early attempts at developing manufacturing tended to be geared to the local, or in some cases the regional market, there were some forays into export-oriented manufacturing in the post-Second World War period. The most prominent examples can be drawn from Mexico, with the Border Industrialisation Programme (Sklair 1993) and Operation Bootstrap in Puerto Rico (Grugel 1995: 172–174; Thomas 1998: Chapter 5). Following the Puerto Rico model, a number of the British Caribbean island states adopted an industrialisation model focusing on export, termed 'industrialisation by invitation', based on the proposals of W. Arthur Lewis, a West Indian economist (Lewis 1950). Foreign investment was encouraged, or 'invited', through tax exemptions, infrastructure provision and changes in tariff regulations. This approach had some success in macro-economic terms; for example, Jamaica's economy grew by up to 8 per cent per annum from 1958–1970 (Grugel 1995: 176), but the focus on urban manufacturing employment exacerbated the wealth divide and the capital-intensive nature of factory production failed to provide the anticipated jobs.

It was only in the 1980s that the region's political economy really shifted towards a more outward-looking, open and export-oriented industrialisation strategy. The globalisation trends of the 1970s and 1980s have clearly had a mammoth impact on the region, particularly in the ways in which manufacturing industry is incorporated into national development strategies, the role of transnational corporations, and the responses of individuals and nation-states to what is sometimes regarded as an unstoppable advance of global capital. Since the late 1970s and especially from the early 1980s, the ISI model has been replaced with an export-orientated one, building on ideas of free trade and foreign direct investment (FDI) as the way forward for indebted nations.

As outlined above, the problems inherent in the ISI model were becoming all too apparent by the 1970s, and the external shocks of oil price rises, interest rate increases and recession in the world economy all dramatically affected the region's economies and prompted some to rethink national development strategies. However, while many governments may have been moving to a more open economy through gradual adjustments, the debt crisis of the 1980s triggered a much more rapid change, usually at the behest of the IMF. A major condition of structural adjustment programmes was the liberalisation of trade, including the removal of tariff barriers and currency devaluations, as well as openness to foreign ownership and investment. The collapse of agricultural markets and prices also pushed many national governments towards an industrialisation strategy.

Multinational and transnational companies seeking to expand their production bases and their markets took this opportunity to set up factories (sometimes called *maquilas* or *maquiladoras*) in the region. Middle America was particularly well placed because of its proximity to the US and the region's low wage costs and relatively well-developed infrastructure. Particular incentives in export-processing zones (EPZs) added to the region's attractiveness. As well as tax exemptions on both profits and machinery, legislation was often enacted to allow for complete profit repatriation, and

Box 6.1 Export-oriented industrialisation in the Dominican Republic

Within Middle America, the Dominican Republic is one of the main sites of export-orientated industrialisation, with over 30 free-trade zones. Manufacturing for export is the third most important source of employment (c. 180,000 workers in 1997) after the public sector and the sugar industry (Safa 1997). Until the early 1980s, agriculture played a very significant role in the Dominican economy, with sugar exports, particularly to the US, dominating export earnings. However, this reliance on one form of revenue generation creates, as in so many other cases, a situation of great vulnerability. When sugar prices fell by 32 per cent in the early 1980s and US sugar quotas were cut, the Dominican economy was left in need of alternative foreign exchange earnings. In addition, the cost of debt repayments increased due to the interest rate rises after 1982, meaning that '48 per cent of all goods and services produced in the country went to service the debt' (Grugel 1995: 192). The national government signed an agreement with the IMF in 1983, committing itself to structural adjustment policies.

The Dominican government strategy of diversification was focused on two main economic sectors: tourism and export-processing. To promote FDI, the number of free-trade zones was increased and benefits expanded. Low wages, particularly for female workers, were another incentive. Until 1992, union activity was, to all intents and purposes, banned in the free-trade zones, meaning workers had to endure low wages and poor working conditions with no channels for legitimate protest. A new Labor Code, passed in 1992, supposedly protects workers' rights to union organisation, but has had little effect (Safa 1997).

Textiles and garment production accounted for three-quarters of firms in EPZs in 1995 (Burns 1995 in Gereffi and Hempel 1996: 22), growing rapidly due to access to the US market under the Caribbean Basin Initiative (see Willis and McIlwaine, this volume). The drive for export-processing has certainly had a positive economic impact in terms of raising foreign currency; for example, between 1981 and 1988 manufacturing exports grew 307.4 per cent to a value of US$502.1 million (Safa 1995: 20). However, this has been at the expense of workers' rights and living standards and has created greater income inequality, as well as maintaining dependence on the US. There are also very limited linkages, either backwards or forwards, to other parts of the economy (Kaplinsky 1993).

Sources: Gereffi and Hempel (1996); Grugel (1995); Kaplinsky (1993); Safa (1995, 1997).

restrictions specifying locally sourced inputs were often lifted (see Clark 1997 for details on the Costa Rican case; see Box 6.1 on the Dominican Republic).

While national governments in the region were often forced to accept greater trade openness in order to obtain further loans or additional aid,

Table 6.2 Statistics on Mexican *maquilas* 1994–2000

	1994	1995	1996	1997	1998	1999	April 2000
No. of maquila plants[1]	2064	2267	2553	2867	3130	3408	3550
Gross production (bn US$)[2]	27.46	41.84	58.80	70.01	86.44	73.63	81.13
Employment[1]	n.a.	681,251	799,347	936,821	1,038,783	1,207,283	1,243,115

Sources: [1] INEGI in Maquila Portal (2000); [2] Ciemez-Vefa in Maquila Portal (2000).
Note
n.a. = not available

it is also clear that the promotion of FDI and export-processing was regarded as a potentially profitable strategy for generating export earnings and creating employment (Klak 1995). Given its relatively long history and locational advantages, the development of the Mexican *maquila* industry is not surprising (see Table 6.2). It generates more foreign exchange than either tourism or petroleum industries, and 20 per cent of Mexican manufacturing jobs are found in *maquila* factories (Maquila Portal 2000). Elsewhere in the region, *maquila* employment is also growing; for example, in July 1999 there were 225 assembly factories (largely in the garment industry) in El Salvador, employing 68,000 workers (National Labor Committee 2000a).

A common outcome of liberalisation policies is the decline in the role of domestic producers unable to compete with the foreign companies attracted to the region. In addition, cheap imports may also undermine home-grown produce and drive domestic companies out of business, or into subcontracting relationships with transnational capital. The garment industry is one of the most successful manufacturing sectors in the region, partly because of the proximity of the massive US market, but also because of the ease and speed with which garments can be assembled. It is also a sector that had a well-developed domestic production base that has been greatly affected by the influx of foreign investment. In 1980, 85 per cent of clothing demand in Jamaica was met by domestic producers. About 5,000 workers were employed in the largely Jamaican-owned garment industry and although 75 per cent of production was for the local market, US$10 million worth of clothing was exported. By 1992, the situation was very different, with only 15 per cent of domestic demand being met by local production, a garment industry workforce numbering 32,000, and locally owned companies contributing less than 20 per cent to total clothing exports. While American companies were greatly involved in this export production, Asian companies (particularly from South Korea and Hong Kong) seeking preferential access to the US market were also very important (Green 1998; Willmore 1994).

As openness and diversification were adopted in most of the region, it became clear that competition within the region was going to work to the advantage of transnational corporations (TNCs) as they could play countries off against each other, as well as having the option of locating elsewhere in the world (for example, East Asia) if Middle America did not meet their needs. Middle American governments adopted the discourse of globalisation to tempt investors to locate in their country. Klak and Myers (1998) provide a fascinating account of how four Caribbean governments (Barbados, St Lucia, Jamaica and Haiti) promote their countries to foreign investors. Despite the diversity of economic and human development in these countries (see McIlwaine, this volume; Willis and McIlwaine, this volume), governments use the same language of efficiency, excellent service provision, cheap labour force, political stability and an attractive environment. As elsewhere in the world, national elites often justify policies which undermine local production, drive down wages and harm workers' working conditions, by referring to the 'unstoppable' nature of globalisation. However, this fails to recognise the ideological nature of globalisation as discourse and the fact that elites can use this idea to further entrench their own local positions of power (Kelly 2000).

Tourism

While manufacturing industry may have been regarded as the 'best' route to development in the past, tourism has also been a key strategy for achieving economic growth, earning foreign currency, and creating employment (Harrison 1992; Lea 1988). In some cases (for example, the Dominican Republic – see Box 6.1), tourism and manufacturing provide a dual-pronged approach to diversification, while for some of the region's economies (especially the smaller Caribbean islands) tourism provides a potential escape from dependence on agriculture.

Advances in air travel have been crucial in developing Middle America as a potential holiday destination not just for North Americans, but for European travellers as well. Many of the Caribbean islands, plus coastal locations on the mainland, have proved attractive for beach holidays, while the Mayan ruins of southern Mexico and Central America are also of great interest for foreign visitors. As tourist preferences change, some of the region's governments have sought to promote niche tourism, such as eco-tourism in Belize and Costa Rica (see Barton, this volume).

Tourism represents a massive source of revenue for many parts of the region. It is, for example, the main source of foreign currency for Cuba (see later) and Jamaica (US$1.12 billion in 1997) (EIU 1998). In some cases, dependence on monoculture has been replaced with over-reliance on tourism, complete with the same seasonality and vulnerability to climatic disasters, as well as changing tastes in the market. Hurricane 'Floyd' hit the Caribbean in September 1999, affecting the tourist industry. Fortunately September and October are traditionally slower months in tourism terms, so the situation could have been much worse. However, there were some losses; for example, the tourism ministry in the Bahamas estimates 'Floyd' led to a reduction in revenue of US$70 million, which represents 5 per cent of estimated tourism expenditure (EIU 1999a).

As well as direct tourism revenue, the development of tourist services is regarded as having indirect effects on the local economy through the tourist multiplier (Mathieson and Wall 1982; Pearce 1981). For example, the construction industry is given a boost through infrastructure and hotel-building projects, and local food producers expand to meet the increased demand for their products. In addition, those earning a living from tourists, either in the formal sector as hotel staff, for example, or as street traders selling handicrafts, spend their money locally, thus supporting other businesses and families. For some of the smaller states of the region, tourism is key to daily survival. In Aruba, approximately 30 per cent of the working population is employed either directly or indirectly in the tourism sector (EIU 1999a).

However, this virtuous circle does not always operate to the benefit of the local or regional economy. The expansion of the tourist industry has been associated with the leakage of profits out of the region to the head-quarters of international hotel chains and travel companies, and in many cases linkages to other sectors of the local economy, such as agriculture, are limited, as food, furniture and fittings are imported from overseas to meet the perceived tastes of the international traveller (see Momsen 1998). This isolation is often exacerbated by the development of all-inclusive private resorts (Pattullo 1996: Chapter 4).

In addition to these potential negative side-effects of tourism development in economic terms, it is clear that the region's tourism boom has had serious environmental and cultural impacts. Although the example of earlier developments, such as Acapulco in Mexico, has flagged up the potentially disastrous impact of rapid and loosely controlled urban development, many governments, often encouraged by the pressures for neoliberal reform, have chosen to sacrifice environmental protection for short-term economic gain. The explosion of the resort of Cancún, on Mexico's Caribbean coast, is a case in point. While planners have attempted to control the resort's development because of the fragile local ecosystem, there are many incidents of environmental pollution and a disregard for controls on unsuitable development (Barton, this volume; Simon 1997).

It is also unsurprising that the influx of millions of tourists a year has a major impact on the societies of the region and cultural norms and practices. The spatial and social separation of tourists from local residents in many of the region's tourist centres has created a situation of mistrust and resentment. Pattullo (1996: 80–1) discusses the lyrics of 'Alien', a St Lucian calypso with the chorus line 'Like an alien, in we own land', referring to the privatisation (often illegally) of beaches and the use of security guards to 'protect' the tourists. Many commentators also highlight the ways in which tourism is linked to increasing crime, the sex industry, gambling and drug trafficking (Harrison 1994; Pettman 1999).

Financial services

The development of financial services industries has been another import-ant dimension of the economic diversification of the region, and has been linked to the deregulation of financial markets and technological change (Held *et al.* 1999). These changes included not only adaptations within existing financial centres because of innovations in financial products, but also the development of new centres, often in marginal locations. Offshore financial centres (OFCs), of which there are several in the Caribbean, rep-resent the spatial outcome of some of these changes dating from the 1970s onwards.

Roberts (1995: 239) has termed the OFCs of the Caribbean 'entrepre-neurial islands', and has outlined the ways in which the success of the Cayman Islands in taking advantage of changes in the global financial system has been mimicked by other countries in the region, with varying degrees of success. Other OFCs include the Bahamas, Bermuda, Panama and the British Virgin Islands. The proliferation of such centres is a reflec-tion of both supply and demand factors (Roberts 1995). Centres specialise in different products and services, and investors are usually careful to spread their risk across a number of centres. In supply terms, financial services offer small, resource-limited nations and territories the chance to gain foreign currency (see Box 6.2).

Technological advances in e-commerce may also be used to promote economic growth in the region; for example the Bermudan government is planning to promote the island as a centre for e-commerce, building on its status as an OFC. The establishment of the Ministry of Telecommunica-tions and E-Commerce is a clear indication of the government's plans, although success will depend partly on legislation and mechanisms to ensure data security (EIU 1999a). The Jamaican government has also pro-moted Jamaica as an ideal location for data-processing services due to the English-language skills and communications networks in the country (Mullings 1998; Willmore 1994).

The feminisation of the labour force: women as victims or beneficiaries of economic globalisation?

One aspect of economic diversification within the region that has attracted much attention is the large-scale employment of women, particularly in multinational corporation (MNC) factories. This pattern is observed in other parts of the world where export-oriented plants have been set up, particu-larly in East Asia. In addition, the expansion of tourism and other service industries has provided jobs for women (Chant 1992; Momsen 1994).

Crude figures seem to demonstrate the numerical dominance of women in export-processing factories. These patterns are regarded as worthy of attention as they seem to challenge prevailing norms regarding a 'male breadwinner' and a 'female homemaker' (see Chant, this volume). Given the common assumption that participation in paid employment is a route to 'liberation' or 'empowerment', women's access to waged work in such

---

Box 6.2   Offshore financial services in the Cayman Islands

The Cayman Islands were one of the first locations of an offshore financial centre in the Caribbean. A British Crown Colony, the Cayman Islands are similar to many states in the region, having few natural resources on which to build economic growth. In the absence of large-scale agriculture, mining or manufacturing, the territory has focused its economic development on tourism and finance.

Tourism was developed first, but was insufficient to provide employment for the large number of unemployed Caymanians. In the 1960s, therefore, expansion of the pre-existing banking sector was proposed as a supplementary development strategy. The Cayman Islands were able to build on their position as a tax haven, as well as promoting their location near the eastern US, their excellent communications facilities and their political stability. A number of companies moved from the Bahamas at this time because of worries about the political situation following Bahamian independence in 1973.

This strategy has been highly successful; in 1964 there were only two banks and very few offshore companies, but this had risen to 218 banks and trust companies in 1977 and 13,600 companies and 360 banks in 1985. When a company registers the government receives a fee. In addition to company registration, other financial services include banking and insurance. New forms of financial product and changes in financial regulation have meant that offshore operations are now possible and highly profitable. In many cases these financial companies exist only as 'brass plate' banks.

Despite the financial rewards of the development of the Cayman Islands as an offshore financial centre (OFC), as a form of economic diversification, it suffers from extreme vulnerability to external shocks. In addition there is competition from other OFCs in the region, able to offer the same proximity to the US.

Sources: Roberts (1995); Thomas (1988).

---

great numbers is interpreted by some as a very positive development which gives women greater decision-making power within patriarchal societies. However, this interpretation has been challenged by many who view the feminisation of factory labour as the result of capital's search for greater profits. Women in this case are employed because they can be exploited more effectively than men due to prevailing gender relations. In her work on Mexican factory employment, Kopinak (1995) terms these two opposing camps the 'apologists' (i.e. those who adop a more positive view of female MNC employment) and 'critics' (i.e. those who regard female MNC employment as double exploitation by both capitalism and patriarchy). The problem with such a dualistic framework is obviously its rigidity and inability to incorporate the diversity within the manufacturing sector and the heterogeneity of the workforce, as well as recognising the way in which global processes are played out in particular local cultural, political and economic contexts.

Plate 6.1 Female factory workers, Guatemala City
Photo: Cathy McIlwaine

When assessing the characteristics of a labour force, it is crucial to recognise that they are the outcome of the intersection between employers'
labour demands and the availability of a particular kind of labour. It has
been argued that women numerically dominate the factory labour forces
throughout the region because employers construct an 'ideal' assembly-
line worker; a hard-working, nimble-fingered, reliable and cheap individual. Women, particularly young single women, are often selected, as
employers perceive them as having these appropriate characteristics (Elson
and Pearson 1981; Fernández-Kelly 1983). In addition, women have been
available for paid employment in greater numbers than in the post-Second
World War period because of economic necessity as male wages plummet
and male unemployment rises in the wake of economic crisis.

Women's entry into the labour force is often regarded as a negative
process because it entails the exploitation of women who are paid far less
than their male counterparts (Kopinak 1995). In addition, researchers
have often highlighted the poor working conditions, and the sexual
harassment that women experience in the workplace (Dwyer 1994). Rather
than being empowered, it is argued, women find themselves exhausted
after a day's paid work combined with the need to complete domestic
chores, as women's entry into paid work is often not mirrored by men's
greater involvement in housework.

Other researchers have gone beyond this homogeneous view of female
employment and have recognised the differences between women, and how
the new employment opportunities may be interpreted and experienced

146

Box 6.3 Women's employment in the Mexican *maquiladoras*

Although the Border Industrialisation Programme was set up to provide work for Mexican men who had previously worked as agricultural labourers in the US under the Bracero Programme, the workforce in the border factories was, from the start, dominated by women. However, while this numerical dominance continues, there are sectoral differences, with women particularly concentrated in clothing and textiles, electronics and toys and sports goods, while male operators outnumber their female counterparts in chemical factories, transport equipment and the manufacture of machine tools. Of course, there is also very obvious vertical segmentation of the labour force, with the upper echelons being dominated by men, and women being concentrated in the lowest paid, and least secure jobs (Pearson 1995; Peña 1997). For young single migrants, accommodation may be provided by the company. While this may be welcomed by recent arrivals in the city and can provide a useful support system, the ways in which corporations use the dormitory system to pervade the 'private' lives of their workers may be problematic (Cravey 1998).

While much of the work on women in MNC factories stresses the levels of exploitation, factory work does present some women with opportunities to advance. Wright (1997) provides a fascinating case study of Gloria, a Mexican woman who skilfully uses her position as a pivot between the Mexican factory-floor workers and the American (often Mexican-American) managers. She works her way up the corporate ladder to become a key member of the management team, despite her refusal to learn English or to wear 'appropriate' clothing. This may be an unusual case, but it does provide some balance to the stereotypical views of women in MNC employment.

An interesting development which has been highlighted in Ciudad Juarez by Cantazarite and Strober (1993) is the increased employment of men in assembly-line work. This suggests that not only are employers willing to take on men, despite the supposed advantages of women, but also that men are willing to work in what have been, until recently, 'women's jobs'. Cantanzarite and Strober argue that this is a response to macro-economic changes that have altered the 'relative attractiveness' of male workers for factory owners, and factory jobs for unemployed men.

Sources: Cantanzarite and Strober (1993); Cravey (1998); Pearson (1995); Peña (1997); Wright (1997).

in a positive way. Discussing female factory employment at a global level, Lim (1990) calls for this 'more realistic' approach to research, criticising the overly negative conclusions that are often made. It is clear that for some women within the region, access to employment has increased their self-esteem, improved their standards of living, and in some cases has enabled them to leave an abusive relationship to set up home alone with their children (Chant 1991, this volume; Safa 1995) (see Box 6.3 on Mexico).

Workers' rights in a globalising world

In the drive to attract foreign investment, governments often highlight the passive nature of their workforces. As Klak and Myers (1998) show, promotional brochures rarely detail the anti-union legislation enacted in many countries since the early 1980s, or provide details of state action against workers' organisations.

Low wages are a key factor in attracting FDI to parts of the region, prompting some commentators to refer to governments' efforts to attract foreign investment as a 'race to the bottom'. It is clear that for many workers in the region, the struggle to make a living has become more and more difficult due to currency devaluations reducing the wage levels in real terms, and increasing prices of consumer goods due to import costs and cuts in government subsidies. However, using wage figures alone can be misleading, because it does not take into account the cost of living in a particular location. The National Labor Committee (2000b) uses the example of a garment worker sewing Wal-Mart clothing in Honduras to demonstrate that factory wages are often so low that households barely manage to scrape by. On a daily wage of US$3.47, a female factory worker with four dependants would be in debt after she had paid for travel to work, a very basic diet (beans, rice, tortillas, potatoes, pasta, eggs, coffee) and rent for one room in a dangerous neighbourhood. Money for education, medicines, clothes and utilities is not available, so families are trapped in poverty.

As well as low wages, workers often have to endure poor working conditions, including long working hours, poor working environments and harsh treatment. Safety regulations are sometimes flouted as companies seek to maximise profits. Dwyer (1994) describes how workers in *maquilas* in northern Mexico often work with chemicals without protection from gloves or masks, often in poorly ventilated rooms.

Throughout the region labour organisation is limited for a variety of reasons. In some countries, such as El Salvador, union activity is banned in assembly factories. The National Labor Committee (2000a) outlines how garment workers in the Korean-owned Doall factories in El Salvador were dismissed for challenging working conditions, or because they were suspected of organising. This action goes against the code of conduct adopted by the Liz Claiborne clothing company, which contracted Doall to produce garments for export to North America.

Another reason for limited union activity is the move in some sectors towards 'workers' councils', modelled on the Asian style of industrial organisation. This model is found particularly among the Asian garment companies. Green (1998) highlights how, in the English-speaking Caribbean, this model is an extreme contrast to the more confrontational style of 'traditional' union activity which is still found in some sectors, such as the Jamaican bauxite industry. Her work in Jamaica suggests that the women factory workers may be more open to this style of employer–employee relationship, while many labour organisers are adopting the neoliberal discourse to support links between increased productivity and wage rises.

Scaling up political protest, particularly by linking into fair trade campaigns, can make a difference, as companies are often sensitive to the feelings of consumers, rather than the demands of their labour force. Many clothing and sports equipment companies have adopted codes of conduct to highlight their commitment to fair and decent working conditions in their factories, but there are numerous examples of these codes being flouted. In addition, consumer pressure may lead to the closure of a factory, or the end of a subcontracting arrangement as transnational companies move production elsewhere. Following consumer complaints about the treatment of factory workers in El Salvador, The Gap pulled out of its subcontracting agreements there in December 1995 (Figueroa 1996).

In the absence of formal forms of organising, workers may still engage in everyday forms of resistance, which both relieve the pressure, albeit temporarily, of the daily grind, and also demonstrate and reinforce the agency (to a limited degree) of factory employees. Peña (1997: 9) is keen to stress that 'the working women of Mexico's *maquilas* are not quite the quiescent victims of managerial fantasies and scholarly theories'. He describes how workers use strategies such as sabotaging equipment and working very slowly as forms of resistance.

Cuba: the exception to the rule?

The economic diversification into export-oriented manufacturing, tourism and financial services, has been embedded in the move towards neoliberalism in the region, with structural adjustment policies adopted as tools to create economies that are more open (some would say exposed) to external forces. However, it is worth remembering that the adoption of World Bank-supported policies has not been universal, and that throughout the twentieth century a number of countries in the region attempted to adopt an alternative route to development other than the capitalist road proposed by Northern governments. Michael Manley's adoption of 'democratic socialism' in 1970s Jamaica and Forbes Burnham's 'co-operative socialism' in Guyana are good examples (Thomas 1988), as is the ten-year reign of the Sandinistas in Nicaragua. However, it is only Fidel Castro's regime in Cuba that can currently be regarded as following anything other than a neoliberal path in the region.

Because of this unusual position, it is worth dwelling on the nature of the Cuban economy and the degree to which the economic diversification processes described above are also evident in Cuba. At the time of the revolution in 1959, sugar constituted 80 per cent of export earnings, and this external dependency continued after 1959. During the 1960s, the Cuban economic development strategy focused on ISI in an attempt (as elsewhere in the region) to reduce dependence on agricultural exports. While there were some successes, overall the industrialisation drive failed due to the pre-existing limitations of a small domestic market, as well as difficulties in accessing inputs because of the US trade embargo and the problems associated with shifting production to Soviet-manufactured machinery. The focus on industry led to a relative neglect in agriculture, so when industry failed to take off, the sugar earnings were unable to fill the gap (Grugel 1995: 206–207).

Manufacturing industry continued to be important in the 1970s and 1980s, but the sugar industry remained key, especially because of the sugar-for-oil trade arrangements with the Soviet bloc. Industry was regarded as complementary to agriculture, with sugar-processing and the production of agricultural machinery dominating (Grugel 1995). There was some diversification in exports but the dominance of sugar remained; between 1980 and 1986 sugar and sugar by-products fell from 83.7 per cent of exports to 77 per cent (Safa 1995: 29).

The collapse of the Soviet bloc has clearly had severe impacts on the Cuban economy. Having gone from a position of dependency on the US (66 per cent of trade in the late 1950s) (Safa 1995: 31), to a subordinate relationship with the USSR and eastern Europe (80 per cent of trade in 1989) (Monreal 1999: 21), Cuba has been left with few economic allies in the post-Cold War period. Collapses in world sugar prices added to the Cuban woes, as did the continued US trade embargo.

In response to the changing geopolitical and economic conditions at the end of the 1980s, in 1990 the Cuban government declared a 'Special Period in Peace Time'. Many of the measures adopted as part of this strategy are very similar to those endorsed as part of structural adjustment policies elsewhere in the region. The state is reducing its role in the economy by encouraging self-employment, including the opening of farmers' markets to allow agricultural producers to sell their produce directly to consumers, rather than to the state (Monreal 1999; Pearson 1997; Susman 1998). There has also been an opening of the national economy to foreign investment, part of what Monreal (1999) calls a 'relinking' to the global economy. In 1997, the first free-trade zones were opened in Cuba; three near Havana and the fourth near Cienfuegos.

Tourism has also been promoted and is now the most dynamic sector in the economy and the main source of foreign currency (US$1.8 billion in 1998) (EIU 1999b). In 1990 there were 340,000 tourist arrivals in Cuba, but by 1998 this had increased to over 1.4 million. (EIU 1999b). While these developments have been welcomed in macro-economic terms, there have been concerns about the 'tourism apartheid' that has resulted, with certain facilities being available only to foreigners (Schwartz 1997).

While it is clear that all these policies have had some success in stabilising the economy, there have been increases in income inequality and levels of poverty. Pearson (1997) provides a fascinating account of how women, in particular, have been affected by the move from reproductive service provision by the state to the household. Since there have been cutbacks in health and education services, as well as the provision of care for children and the elderly, women have had to shoulder this burden, at the same time as expanding their involvement in the growing tourism and factory sectors.

Regional co-operation

Regional trading agreements have often been held up as a way in which small, less economically developed nations can compete on world markets.

Middle America is no exception, and the twentieth century witnessed a number of attempts by regional governments to co-operate, particularly in the context of trade. As part of the ECLA development ISI strategy outlined above, Latin American countries were encouraged to work towards regional integration to aid industrialisation (Bulmer-Thomas 1988; Montecinos 1996; Orantes 1972). Within Latin America as a whole, the Latin American Free Trade Area (LAFTA) was promoted, although within Middle America only Mexico was included, and the attempt at such large-scale co-operation failed to take off.

The Central American Common Market (CACM) was one of the earliest attempts, associated with the ISI period. The CACM was created by Costa Rica, Guatemala and Nicaragua in 1963, with Honduras and El Salvador joining soon after. In order to protect and promote newly developed industries, CACM involved the harmonisation of the tariff structure so that imports from countries outside were taxed to the same degree regardless of the destination of the imports. Tariffs were highest on consumer goods, reflecting the priorities of ISI policies and the ease with which particular sectors of imports could be replaced by domestically produced goods (Bulmer-Thomas 1988). Within CACM, tariffs were removed on certain, particularly industrial, goods.

While assessing the success of regional integration is a highly complex procedure, there is agreement that during the 1960s CACM did contribute to increasing industrialisation and growing intraregional trade; for example intraregional trade within CACM, as a proportion of CACM members' total trade, grew from 6.5 per cent in 1960 to 26.1 per cent by 1971 (Grugel 1995: 178). Some of this growth was attributed to the role of foreign investment, as MNCs increased their involvement in Central America, particularly in sectors such as tyres, glassware and chemicals (Bulmer-Thomas 1988; Weaver 1994: 165–173).

However, regional integration requires both political and economic co-operation, and does not fare well when there are perceived or actual inequalities between participants. Despite the setting up of the Central American Bank for Economic Integration to help distribute the financial rewards of regional co-operation, and 'integration industry provisions' to encourage firms to set up in locations to balance regional inequalities, wide differences persisted, with El Salvador in particular benefiting from the new industries. Following the Honduran–Salvadoran War (the 'Soccer War') of 1969, Honduras withdrew from CACM. This, combined with the growing limits on ISI potential and then later political unrest in the region, the debt crisis and adjustment policies, led to a slump in intraregional trade and a stagnation of the workings of CACM (Bulmer-Thomas 1988).

The Caribbean Common Market (CARICOM) is a similar enterprise seeking to overcome the limited resources and small domestic markets of the Caribbean nations, to promote regional economic development. This grouping developed out of the Caribbean Free Trade Association (CARIFTA), which began in 1965 as an agreement between Antigua, Barbados and Guyana, but was expanded in 1968 to include all ten members of the West Indies Federation and Guyana. While CARIFTA represented an attempt at regional co-operation, it was at the basic level of reduced trade restrictions within the region. There was no attempt to implement common external tariffs, or to promote greater freedom of movement of either capital or labour (Thomas 1988).

The grouping changed to CARICOM in 1973 and has welcomed new members, such as Belize in 1974 and the Bahamas in 1983. CARICOM was promoted as attempting greater regional integration through the implementation of a common external tariff, for example. The CARICOM agreements also included co-operation in foreign policy and arrangements for *ad hoc* co-operation on social issues (Thomas 1988). However, only the Organisation of Eastern Caribbean States (OECS), set up in 1968, has made significant progress towards a common market. The OECS is a member of CARICOM and consists of the Leeward and Windward Islands. There is a common external tariff, as well as a single currency (the EC dollar) and a central bank (Green 1995).

Intraregional trade was largely based on agricultural goods (rice was the most widely traded agricultural product), and oil from Trinidad and Tobago. This pattern is unsurprising, given the economic profile of the nations involved. However, there was also an increase in manufactured goods being traded, reflecting both existing industries in the Caribbean and newly developed industries as part of ISI policies. However, between 1973 and 1985, intraregional trade never rose above 10 per cent of the total trade, and the figure was dominated by the oil exports (Thomas 1988). As with CACM, attempts by CARICOM's member states to enhance national development through regional co-operation experienced problems during the 1980s due to declines in trade as adjustment policies were implemented (Thomas 1988: 312–315).

One of the features of recent globalisation processes has been the rise of regional blocs in both political and economic terms. Three economic blocs now dominate the world economy, and are often referred to as the 'triad': Europe, North America and East Asia (focusing on Japan). Given the overwhelming economic and political power of these groupings, how can the nations of Middle America hope to compete? Given the past record of limited success in the field of regional economic co-operation, is such an approach feasible, or should the Central American and Caribbean nations follow Mexico into an agreement with the US and Canada? Montecinos (1996) highlights how the debt crisis has led to a rethinking of development strategies in Latin America, and how regional integration is a key part of the consensus regarding development strategies. However, there is a recognition that, rather than providing protective barriers behind which governments can implement national development policies, regional integration should be seen as a platform from which to launch into the global economic system, i.e. regional integration is regarded as a key part of export-oriented free-trade strategies.

The Mexican path into formal trade agreements with the US and Canada is regarded as a model for many of the region's nations. As Green (1995: 146) highlights, it is very different from previous regional trade agreements in the Americas as it involves co-operation between 'developed' and 'developing world' countries. After long negotiations NAFTA came into force on 1 January 1994. Under the Agreement, restrictions on trade between the three nations are to be lifted over a 15-year period, although there are exceptions relating-to particular sectors. The free movement of labour is not included within the arrangements. It is clear that NAFTA has increased the dependence of Mexico on the US as a market for exports, but NAFTA has also promoted industrial development, particularly on the northern Mexican border, through investment by TNCs (Weintraub 1996). However, there are also a multitude of problems associated with NAFTA, in addition to the environmental and social issues highlighted in the earlier section on *maquilas*. Free trade means greater competition for domestic producers, and within Mexico peasant producers and industries that had been protected through ISI policies have suffered because of cheaper imports (Green 1995: 152). Regional inequalities have also increased as a result of differential incorporation into the NAFTA process (Morales 1999). The Zapatista uprising in Chiapas (see Howard, this volume) is linked to the perceived negative outcomes of NAFTA. Marshall (1998) highlights some of the concerns held by Caribbean governments regarding Mexico's incorporation into NAFTA. Despite these potential problems with greater freedom of trade with regional associations, many of Middle America's governments are lobbying for inclusion in such organisations (Watson 1996).

Conclusions

This chapter has outlined the non-agricultural economic development strategies that have been adopted in the region in recent years. What is clear is that although export-processing industrialisation, financial services and tourism are the key sectors, the balance between them and their success vary greatly depending on natural resource endowments, local political conditions, and characteristics of the labour force. It is also clear that success can be transitory as the region's economies are vulnerable to changes in market taste, climatic conditions and global economic patterns. Diversification helps reduce this vulnerability, but for the small states of the region, autonomy and influence at an international level are extremely limited. This does not mean that the governments and peoples of the region are passive, as demonstrated by the attempts at regional co-operation, but options are limited and forms of resistance may be through 'non-political' channels.

Useful websites

www.acs-aec.org Association of Caribbean States. Useful information about regional co-operation, national statistics and sector reports.

www.maquilasolidarity.org Maquila Solidarity Network. Canadian-based organisation reporting on factory conditions in Latin America, the Caribbean and Asia. Useful up-to-date reports on campaigns and labour disputes.

www.nlcnet.org National Labor Committee. A US-based human rights organisation seeking to promote workers' rights in factories throughout the world.

Further reading

Figueroa, H. (1996) In the name of fashion: exploitation in the garment industry, *NACLA Report on the Americas* **XXXIX**(4): 34–40.
A brief but fascinating overview of the way in which transnational clothing companies operate within the region.

Klak, T. (ed.) (1998) *Globalization and Neoliberalism: the Caribbean Context.* Rowman & Littlefield, Oxford.
An excellent collection of chapters discussing the way in which globalisation has been experienced in the Caribbean region.

Peña, D. (1997) *The Terror of the Machine: Technology, Work, Gender and Ecology on the US–Mexico Border.* CMAS Books, Austin, TX.
An engrossing discussion of Mexican *maquila* production based on fieldwork with factory workers since the early 1980s.

Pattullo, P. (1996) *Last Resorts: the Cost of Tourism in the Caribbean.* Cassell & Latin American Bureau, London.
An accessible and wide-ranging discussion of the nature of tourism in the Caribbean.

References

Bulmer-Thomas, V. (1987) *The Political Economy of Central America since 1920.* Cambridge University Press, Cambridge.

—— (1988) The Central American Common Market, in El-Agraa, Ali M. (ed.) *International Economic Integration*, 2nd edn. Macmillan, Basingstoke: 284–313.

Cantanzarite, L. and **Strober, M.** (1993) The gender recomposition of the maquiladora workforce in Ciudad Juárez, *Industrial Relations* **32**(1): 133–147.

Chant, S. (1991) *Women and Survival in Mexican Cities: Perspectives on Gender, Labour Markets and Low-Income Households.* Manchester University Press, Manchester.

—— (1992) Tourism in Latin America: perspectives from Mexico and Costa Rica, in D. Harrison (ed.) *Tourism in the Less Developed Countries*. Belhaven, London: 85–101.

Clark, M.A. (1997) Transnational alliances and development policy: Nontraditional export promotion in Costa Rica, *Latin American Research Review* **32**(7): 71–97.

Cravey, A. (1998) *Women and Work in Mexico's Maquiladoras.* Rowman and Littlefield, Oxford.

Dwyer, A. (1994) *On the Line: Life on the US–Mexican Border.* Latin American Bureau, London.

Economist Intelligence Unit (EIU) (1998) *Country Profile: Jamaica.* EIU, London.

—— (1999a) *Country Report: Bahamas Barbados, Bermuda, British Virgin Islands, Netherlands Antilles, Aruba, Turks & Caicos Islands, Cayman Islands. 4th Quarter 1999.* EIU, London.

—— (1999b) *Country Profile: Cuba 1999–2000.* EIU, London.

Elson, D. and **Pearson, R.** (1981) The subordination of women and the internationalization of factory production, in Young, K., Wolkowitz, C. and McCullagh, R. (eds) *Of Marriage and the Market: Women's Subordination Internationally and its Lessons.* Routledge, London: 18–40.

Fernández-Kelly, P. (1983) *For We Are Sold, I and My People: Women and Industry in Mexico's Frontier.* State University of New York Press, Albany.

Figueroa, H. (1996) In the name of fashion: exploitation in the garment industry, *NACLA Report on the Americas* **XXXIX**(4): 34–40.

Gereffi, G. and **Hempel, L.** (1996) Latin America in the global economy: running fast to stay in place, *NACLA Report on the Americas* **XXXIX**(4): 18–29.

Green, C. (1998) The Asian connection: the US-Caribbean apparel circuit and a new model of industrial relations, *Latin American Research Review* **33**(3): 7–47.

Green, D. (1995) *Silent Revolution: the Rise of Market Economics in Latin America.* Latin American Bureau, London.

Grugel, J. (1995) *Politics and Development in the Caribbean Basin.* Macmillan, Basingstoke.

Harrison, D. (ed.) (1992) *Tourism and the Less Developed Countries.* Belhaven, London.

—— (1994) Tourism, capitalism and development in less developed countries, in Sklair, L. (ed.) *Capitalism and Development.* Routledge, London: 232–257.

Held, D., McGrew, A., Goldblatt, D. and **Perraton, J.** (1999) *Global Transformations: Politics, Economic and Culture.* Polity Press, Cambridge.

Kaplinsky, R. (1993) Export processing zones in the Dominican Republic: transforming manufactures into commodities, *World Development* **21**(11): 1851–1865.

Kelly, P. (2000) *Landscapes of Globalisation.* Routledge, London.

Klak, T. (1995) A framework for studying Caribbean industrial policy, *Economic Geography* **71**(2): 297–316.

Klak, T. and **Myers, G.** (1998) How states sell their countries and their people, in Klak, T. (ed.) *Globalization and Neo Liberalism: the Caribbean Context.* Rowman & Littlefield, Oxford: 87–109.

Kopinak, K. (1995) Gender as a vehicle for the subordination of women maquiladora workers in Mexico, *Latin American Perspectives* **22**(1): 10–29.

Lea, J. (1988) *Tourism and Development in the Third World*. Routledge, London.

Lewis, W.A. (1950) The industrialisation of the British West Indies, *Caribbean Economic Review* **2**(1).

Lim, L. (1990) Women's work in export factories: the politics of a cause, in Tinker, I. (ed.) *Persistent Inequalities: Women and World Development*, Oxford University Press, Oxford: 101–119.

Lustig, N. (1992) *Mexico: the Remaking of an Economy*. The Brookings Institution, Washington, D.C.

Maquila Portal (2000) *www.maquilaportal.com/*. Accessed July 2000.

Marshal, D.D. (1998) NAFTA/FTAA and the new articulations in the Americas: seizing structural opportunities, *Third World Quarterly* **19**(4): 673–700.

Mathieson, A. and **Wall, G.** (1982) *Tourism: Economic, Physical and Social Impacts*. Longman, Essex.

Momsen, J. (1994) Tourism, gender and development in the Caribbean, in Kinnaird, V. and Hall, D. (eds) *Tourism: Gender Perspectives*. Wiley, London: 106–120.

—— (1998) Caribbean tourism and agriculture: new linkages in the global era?, in Klak, T. (ed.) *Globalization and Neoliberalism: the Caribbean Context*. Rowman & Littlefield, Oxford: 115–134.

Monreal, P. (1999) Sea changes: the new Cuban economy, *NACLA Report on the Americas* **XXXII**(5): 21–29.

Montecinos, V. (1996) Ceremonial regionalism, institutions and integration in the Americas, *Studies in Comparative International Development* **31**(2): 110–123.

Morales, I. (1999) NAFTA: The institutionalisation of economic openness and the configuration of Mexican geo-economic spaces, *Third World Quarterly* **20**(5): 971–994.

Moser, C. and **McIlwaine, C.** (2001) *Violence in a Post-Conflict Context: Urban Poor Perceptions from Guatemala*. World Bank, Washington, D.C.

Mullings, B. (1998) Jamaica's information processing services: neoliberal niche or structural limitation?, in Klak, T. (ed.) *Globalization and Neoliberalism: the Caribbean Context*. Rowman & Littlefield, Oxford: 135–154.

National Labor Committee (2000a) 'Fired for crying to the gringos.' *www.nlcnet.org/LIZ/FIRED/lizfired.htm*. Accessed July 2000.

—— (2000b) 'Wal-Mart sweatshops in Honduras', *www.nlcnet.org/walmart/honwal.htm*. Accessed July 2000.

Orantes, I. (1972) *Regional Integration in Central America.* Lexington Books, Lexington, Mass.

Patullo, P. (1996) *Last Resorts: the Cost of Tourism in the Caribbean.* Cassell & Latin American Bureau, London.

Pearce, D. (1981) *Tourism Development.* Longman, Harlow, Essex.

Pearson, R. (1995) Male bias and women's work in Mexico's border industries, in D. Elson (ed.) *Male Bias in the Development Process.* Manchester University Press, Manchester: 133–163.

—— (1997) Renegotiating the reproductive bargain: gender analysis of economic transition in Cuba in the 1990s, *Development and Change* 28: 671–705.

Peña, D. (1997) *The Terror of the Machine: Technology, Work, Gender and Ecology on the US–Mexico Border.* CMAS Books, Austin, TX.

Pettman, J.J. (1999) Sex tourism: the complexities of power, in Skelton, T. and Allen, T. (eds) *Culture and Global Change.* Routledge, London: 109–116.

Roberts, S. (1995) Small place, big money: the Cayman Islands and the international financial system, *Economic Geography* **71**(2): 237–256.

Safa, H. (1995) *The Myth of the Male Breadwinner.* Westview, Oxford and Boulder, CO.

—— (1997) Where the big fish eat the little fish: women's work in free-trade zones, *NACLA Report on the Americas* **XXX**(5): 31–36.

Schwartz, R. (1997) *Pleasure Island: Tourism and Temptation in Cuba.* University of Nebraska Press, Nebraska.

Simon, J. (1997) *Endangered Mexico: an Environment on the Edge.* Latin American Bureau, London.

Sklair, L. (1993) *Assembling for Development: the Maquila Industry in Mexico and the United States.* Center for US-Mexican Studies, San Diego.

Susman, P. (1998) Cuban socialism in crisis, in Klak, T. (ed.) *Globalization and Neoliberalism: the Caribbean Context,* Rowman & Littlefield, Oxford: 179–208.

Thomas, C. (1988) *The Poor and the Powerless: Economic Policy and Change in the Caribbean.* Latin American Bureau, London.

UNDP (1998) *Human Development Report 1998.* Oxford University Press, Oxford and New York.

—— (1999) *Human Development Report 1999.* Oxford University Press, Oxford.

Watson, H. (1996) Globalization, new regionalization, and NAFTA: Implications for the signatories and the Caribbean, *Caribbean Studies* **29**(1): 5–48.

Weaver, F.S. (1994) *Inside the Volcano: the History and Political Economy of Central America.* Westview Press, Oxford.

Weintraub, S. (1996) The meaning of NAFTA seen from the United States, in Whiting Jr, V.R. (ed.) *Regionalization in the World Economy: NAFTA, the Americas and the Asia Pacific.* Macmillan India, Delhi: 65–87.

Willmore, L. (1994) Export processing in the Caribbean: the Jamaican experience, *CEPAL Review* 52: 91–104.

World Bank (2000) *World Development Report 1999/2000,* OUP, Oxford and New York.

Wright, M. (1997) Crossing the factory frontier: gender, place and power in the Mexican maquiladora, *Antipode* **29**(3): 278–302.

CHAPTER 7

Sowing the seeds of modernity: the insertion of agriculture into the global market

Andy Thorpe and Elizabeth Bennett

Introduction

Throughout the region agriculture has played a crucial role in economic development, and remains a key sector in many countries. It is an important component of GDP, contributing around one-third in Nicaragua and Haiti. It is also a meaningful source of employment in a number of countries, sustaining 46 per cent of the economically active population in Guatemala, 62 per cent in Haiti and 23 per cent in St Vincent and the Grenadines. Furthermore, it is the cornerstone of many countries' foreign trade, currently accounting for 83 per cent of Honduran, 72 per cent of Nicaraguan and 52 per cent of Costa Rica's exports (see Table 7.1). This dependence is not new: over several hundred years agriculture has consistently been the focus of conscious and unconscious efforts to insert the countries further into global markets. However, it is both the historical and contemporary importance of export-led agriculture to the growth and development of the national economies of the region that warrants a more detailed analysis of the sector.

In order to analyse how the process of globalisation has affected agriculture and agribusiness in the region, this chapter looks at the origins of, the technology behind, and the subsequent development of several key regional crops. Whilst recognising that there are substantial differences between the development of the so-called 'Banana Republics' of Central America and the 'Sugar Dependencies' of the Caribbean, the chapter shows that, in both instances, climatic and topographical factors were highly influential in determining where such enterprises were initially sited. In addition, the process of colonialism, the impact of the US as a major market for the region's agricultural produce and the changing nature of global agribusiness have all had key influences on the agriculture sector.

Table 7.1 Agriculture as a source of regional employment, export earnings and contributor to GDP

Country	Percentage of population engaged in agriculture[1]		Agriculture as a % of GDP[2]	Agriculture as a % of export portfolio[2]
	2000	2050		
Costa Rica	20	5	15	52.8
El Salvador	29	9	12.2	33.4
Guatemala	46	28	20.6	47.2
Honduras	32	6	18.5	83.7
Nicaragua	20	3	34.9	71.8
Mexico	21	5	4.5	5.8
Panama	20	5	7.3	63
Barbados	4	1	5.02[3]	14.1
St Lucia	24	14	10.3	n.a.
St Vincent and Grenadines	23	13	11.2	n.a
Trinidad and Tobago	9	3	1.4	1.4
Cuba	14	5	6.7	3.6[4]
Dominican Republic	17	3	16.7	16.3[4]
Jamaica	21	8	9.3	15.5
Haiti	62	34	28.4	35.23[5]

Sources: ECLAC (1999); FAO various years.

Notes

[1] figures are estimates

[2] data are for latest year available (1998) and include hunting, fishing and forestry

[3] 1997

[4] 1988

[5] 1979

n.a. = data not available

In the context of globalisation, debates have focused on a 'new political economy of agriculture' (Whatmore 1995), highlighting how the twentieth century saw a move towards farm production being linked directly into commodity chains, often controlled by international capital.

However, as Watts and Goodman (1997) argue, the agro-food sector cannot be regarded as a globalised industry in the same sense as electronics or cars, for example. As will be highlighted in this chapter, locally specific physical environmental conditions, as well as differences in political and social structures, mean that changes in external factors are experienced differently across the region. Busch and Juska (1997) highlight the import-ance of considering the way changes in the organisation and focus on local agriculture reflect the actions of particular actors operating within place- and time-specific contexts.

While stressing local specificity, it is contended here that export-led agriculture in the region can be viewed broadly as having passed through three phases. The first phase, which effectively locked the region into the global marketplace, saw the introduction and consolidation of what are now regarded as the region's traditional agro-exports; sugar, bananas and coffee. The second phase was a distinctly Central American phenomenon; a sharp increase in disposable income in the US after the Second World War encouraged Central American landowners to raise beef cattle and cotton, government incentives being deployed to facilitate the process of export diversification. However, these first and second phases by and large encouraged the concentration of land into large farms where technical and marketing economies of scale could be realised. Where this concentration was most pronounced, the price was growing landlessness, rising social tension and a 'backlash' against the agro-export development model. The Cuban Revolution pushed agrarian reform to the fore and, whilst not always successful, many land reform laws approved in the region over the last half-century have impacted upon national tenure structures (see Box 7.1).

By the 1990s, however, new voices had emerged in agrarian policy circles, mirroring wider debates about the direction of state development policies (see Willis and McIlwaine, this volume). Agrarian failings were now attributed to excessive state intervention in the agrarian sphere. The prescription was straightforward – agriculture needed to be modernised and the most effective modernising influence was the market. Modernisation, by redressing currency overvaluation and offering a barrage of incentives to stimulate new export products has consequently triggered a third phase – the cultivation of non-traditional agro-exports such as snowpeas, straw-berries, flowers and macadamia nuts. The question is, although such exports strengthen ongoing globalisation trends, who are the real beneficiaries?

Sugar, coffee and bananas: the first phase

The insertion of Middle America into global trade circuits was largely accomplished through the planting and cultivation of three crops – sugar, coffee and bananas – and was a result of European colonial influences in the region. In the Caribbean, the imperial powers installed sugar plantations

---

Box 7.1 Glossary of land-holding terms

Hacienda: A large estate with a good house, as distinct from a farming estab-lishment with basic accommodation for herdsmen.

Plantation: A large estate in subtropical or tropical climates used to grow just one crop (such as coffee, sugar, cotton or bananas). Labour for the plantations originally came from slaves, but is now associated with labourers who live on the estate, often with few rights.

Plantocracy: Government by planters or the name used for planters collectively.

Ejido: A pre-Columbian form of land tenure in which the land is held in common by the whole village/community. This is still the meaning in Honduras and Mexico, whereas in some other countries it merely means an agricultural co-operative; *ejidal* (pertaining to an *ejido*); *ejidatarios* (holder of a share of an *ejido*).

Latifundio: Very large landholdings, giving a small minority of individuals control of a major part of the arable land. Generally defined in Central America as an estate of over 1,000 *manzanas* (709 hectares).

---

underpinned by slave labour. While slavery was abolished in the colonies between 1833 and 1880, sugar has remained the principal subregional ex-port crop to this day. Central American insertion occurred somewhat later, the economic and political disarray that succeeded independence from Spain being arrested only following the introduction of coffee and bananas in the latter half of the century (Bulmer-Thomas 1994: 1). As trade grew, economic prosperity resulted and political strife lessened, albeit at the cost of overdependence upon the export of these two crops (see Table 7.2). Mexico remains rather distinct, particularly after the 1910 revolution, with greater focus on land redistribution for peasant agriculture (see Box 7.2) and developments in the manufacturing sector.

Sugar

Sugar was first cultivated in the region in the early sixteenth century, the first shipment of sugar leaving Santo Domingo in 1516. Early attempts to grow and process the crop failed, however, due to lack of support from Spain, which perceived Caribbean sugar as a threat to the more proximate industry in the Canary Islands. Consequently, the Caribbean's early sugar endeavours faded until the British and the French began production in Barbados and Martinique around 1650 (Mintz 1986: 32) encouraged by an increasing European demand for the crop and the emergence of the 'New World' plantation system. After slaves were freed, labour arrangements on sugar plantations had to adapt to the new circumstances. In the Anglo-Caribbean, many freed slaves moved off plantations and began growing tropical subsistence crops on marginal lands, whilst in Cuba freed slaves stayed within the sugar economy (Ayala 1995: 96). Although the likes of John Stuart Mill were campaigning for these smallholders to be bequeathed

Table 7.2 Export of coffee and bananas, as a percentage of total exports, from selected Central American countries (1929)

Commodity	Costa Rica	El Salvador	Guatemala	Honduras[1]	Nicaragua
Coffee	67.2	92.6	76.6	2.1	54.3
Bananas	25.2	–	12.9	84.9	18.3
Coffee and bananas	92.4	92.6	89.5	87.0	72.6

Source: Bulmer-Thomas (1994: 34).
Note
[1] 1928/9

Box 7.2   The significance of Mexican agrarian reform

The Mexican revolution ushered in the first meaningful Latin American agrarian reform legislation, providing as a consequence both an inspiration for, and yardstick by which to measure, subsequent land redistributions in the region.

Land distribution had become increasingly skewed in the pre-revolutionary period. The intrusion of commercial agriculture into the rural subsistence economy was accentuated following legislation in 1856 (which required the sale of all religious and civic landholdings) and 1883 (which saw land survey companies receiving title to 12.7 million hectares in return for surveying and titling 38 million hectares (Markiewicz 1993: 15)), and precipitated a series of caste wars. Tannenbaum (1968: 28, 54, 92) estimates that by 1910, 83 per cent of the rural workforce were landless, and 59 per cent of those fortunate enough to own land owned less than five hectares. In contrast, 110 ranches each held more than 100,000 hectares. A combination of static money wages and escalating grain prices between 1890 and 1910 triggered a series of peasant revolts (the main ones were led by Emiliano Zapata and Pancho Villa), which succeeded in placing land reform at the forefront of the revolutionary agenda. Venustiano Carranza, a revolutionary leader, won the allegiance of agrarian and labour groups by encapsulating a number of the key agrarian (and urban labour) demands in a 1915 decree, and went on to claim the Presidency. In 1917 the agrarian programme was enshrined in the Mexican Constitution via Article 27.

Article 27 defines, albeit in somewhat general terms, (i) the nature of, and limits to, private property ownership, (ii) who is entitled to own private property, and (iii) procedures for resolving agrarian conflicts. This legislation saw landholdings restored to communities in the form of *ejidos* and provided an avenue by which successive Mexican presidents could advance or frustrate peasant demands for greater land access.

Sources: Markiewicz (1993); Tannenbaum (1968).

land officially, seeing the use of smallholdings as a useful social release valve in many of the more overpopulated islands, the plantocracy did not agree (Richardson 1997: 5).

As with other agricultural crops, technology radically changed not only the way sugar was processed, but also the way it was grown. Prior to the nineteenth century, most sugar cane was grown on small plantations and the finished product was of the muscovado or demerara type. With the advent of the steam-powered sugar centrifuger in the mid-1800s, the production of white sugar, increasingly in demand due to its pleasing appearance, became easier, and the sugar industry changed rapidly. The high capital costs and overheads of the new steam-powered technology required larger plantations to generate sufficient profits to cover costs. However, many of the smaller islands could not expand their plantations any further and these spatial constraints prevented them from adopting the new technology (Richardson 1997).

Of course, those islands able to adjust to the new situation enjoyed high returns. St Lucia had almost ceased its production of muscovado sugar by the 1890s and was instead producing centrifugal white sugar for a large and expanding US market. The plantocracy elsewhere in the British Caribbean was more resistant to change, a conservatism which saw planters in St Vincent, for example, never fully engage with the new technology. The plantocracy in Barbados was equally unwilling to embrace the new, less labour-intensive technology for fear of provoking social conflicts if people were laid off. One island in particular did not suffer unduly. Whereas in 1815 Cuba had been exporting 50 per cent less sugar than Jamaica, by 1894 the country was producing 50 times more than Jamaica and four times more than all the British colonies (Knight 1978: 240).

Cuba increasingly came to dominate the new global market for sugar for a number of reasons. First, slavery was abolished much later in Cuba (1880) than in the rest of the region (1830) so cheap labour was available. Second, sugar production in Cuba became ever more concentrated during the 1880s and 1890s as economies of scale were exploited effectively. For example, although the number of mills in the Remedios region declined from 40 to 17 between 1878 and 1894, sugar production doubled (Ayala 1995: 96).

The plantation system developed where economies of scale and process linkages benefited large-scale production. Because plantation production was destined for export, the system was wholly extractive and relied heavily on external markets in London and other European centres. By the turn of the nineteenth century, Tate and Lyle was the largest company importing Caribbean sugar cane.

In the early 1800s, however, German scientists developed a means of extracting sugar from beet grown in northern Europe. By 1880 beet production equalled cane production – thanks to the introduction of sugar 'bounties' that refunded internal excise taxes on exported beets. The presence of bounties encouraged a sharp increase in beet production across Europe and threatened the world price for cane sugar. Matters were made worse for British Caribbean sugar producers as the preferential purchasing agreement for colonial sugar ceased in 1846 (partly at the insistence of East

Anglian beet farmers). Whereas the small islands had a comparative advantage in the seventeenth and eighteenth centuries, the abolition of protective tariff barriers, the introduction of steam, and the threat of a beet substitute effectively removed any advantage.

By the late nineteenth century many of the traditional sugar economies in the Windward Islands were severely depressed and shifted towards producing subsistence tropical crops. The centre of the sugar industry effectively relocated itself, then, to those islands, such as Cuba, Jamaica, Haiti and the Dominican Republic, able to support larger plantations. It would take more than half a century before the smaller islands of the Caribbean once again found themselves at the centre of a global agricultural market.

The demise of sugar in the Anglophone Caribbean was not wholly due to beet technology. A vital component behind the expansion of production in places such as St Lucia and Cuba can be traced to the emergence of the US as an economic force in the region. A growing industrial base and an increased need for raw materials after the devastation of the American Civil War (1861–65) had drawn the Caribbean basin into the US sphere of influence by the second half of the century. In a bid to maintain political stability in the region and to prevent the further spread of European influence, the US had an interest in developing markets close to home (Randall and Mount 1998). Consequently, following the island's independence in 1902, a series of reciprocity agreements assured the US of a steady supply of sugar whilst keeping Cuba within its influential orbit until the 1959 revolution (Azicri 1988; Martinez-Alier 1977).

Bananas

As with sugar, European influence was of great importance in the development of banana production, as bananas were introduced into the region by the Portuguese in the fifteenth century. However, unlike sugar, bananas were destined solely for the internal market until the American entrepreneur Lorenzo Dow Baker commenced export operations in June 1870. Unfortunately, this nascent trade in bananas was hindered by both the perishability of the product and the lengthy sea voyages to the US. Nevertheless, the potential profits were enormous – a US government publication at the time suggested that, for an outlay of US$8,400 over two years, a Costa Rican planter could expect to recoup US$27,000 (Read 1983: 181). However, few growers at the time had the luxury of contracts with shipping companies, and produced in the hope of being able to sell as the crop matured, while shipping companies would often put into port in the hope of finding fruit ready for exportation.

Through its contacts with Lorenzo Baker, Jamaica was the focal point of this early trade. The Boston Fruit Company, established in 1885 by Baker and a number of associates, saw initial investment appreciate 3,500 per cent within 5 years, the profits being ploughed into the establishment of plantations in Jamaica and subsequently Cuba and the Dominican Republic. Elsewhere, entrepreneurs were also beginning to invest in banana production. In Costa Rica, the trade evolved largely by accident. After the original company contracted to build the railway linking San José to the Caribbean coast at Limón was bankrupted in the early 1880s, the task of completing the railroad was awarded to the American, Minor C. Keith. Cash-flow problems saw Keith attempt to defray his costs by exporting the bananas growing alongside the track to New York and New Orleans. The venture proved so successful that, by 1898, Costa Rica had become the second biggest global exporter of bananas behind Jamaica. In Panama the early trade was also dominated by Americans (the Frank Brothers Co), whereas in Guatemala and Honduras production remained in the hands of small and medium local producers until the early decades of the twentieth century.

As with both sugar and coffee, technological developments were to revolutionise banana production in the region. In the early days of the trade, the export window was highly seasonal. Shippers were reluctant to transport bananas during the summer (the fruit ripened too quickly) or to the northern US ports during the winter (the fruit froze). This changed with the advent of steamships in the latter part of the nineteenth century. Steam power not only resulted in more rapid transit times, so extending the export window, but also encouraged the construction of ever larger vessels so as to benefit from technical scale economies.

It now became imperative, if ships were to depart fully loaded and on time, to ensure continued supplies. Initially, supply was secured through contracts with local growers. However, over time, local growers were supplanted by corporately owned plantations, the early stages of what, in the twentieth century, became much more globalised production processes. This tendency towards vertical integration was facilitated by developments on the communications front. Telegraph and telephones enabled incoming vessels to communicate with the dockside, the dockside in turn communicating with the growers, so as to ensure that fruit was ready to be loaded the moment the vessel docked. The development of transport infrastructure also assisted this expansion, with railroad growth in the US enlarging demand for the fruit, while railroad expansion in Central America allowed greater volumes to be transported more rapidly to the ports.

Central America now came into its own as a banana producer. These states, possessing an abundance of land but suffering from severe shortages of capital, were only too willing to ally with foreign banana companies. In return for building railroads – albeit often serving no purpose other than the interests of the companies themselves – the companies were granted extensive land rights (Thorpe 2000). The Caribbean and, to a lesser extent, the Pacific coastlines were ideal banana-growing territory – flat, humid and easy to irrigate, relatively windless and with good soils – offering the opportunity to realise pronounced production economies of scale through the establishment of large plantations. By 1913, bananas dominated the export portfolios of Costa Rica (50.3 per cent), Honduras (50.1 per cent) and Panama (Bulmer-Thomas 1994: 8). The plantation system has two weaknesses, however. First, there is a very real probability that fungal infections or diseases can knock out entire plantations. Because of this companies historically acquired landholdings across both Central and South America, switching production if outbreaks occurred. The second problem is labour militancy. Large labour concentrations, common within plantation economies, provide a fertile ground for labour organisation. This led to strikes for higher wages across banana plantations in Costa Rica (1934) and Honduras (1920 and 1954). While the companies generally acceded to the workers' demands in the short term, in the longer term this led to labour being laid off as the companies both mechanised production and harvesting processes and increasingly contracted out cultivation to small and medium local growers (Posas and Del-Cid 1983: 145).

Unlike either the coffee or sugar trade, which were originally geared towards the European market, the regional banana trade from its inception was oriented to meeting US demand. Consequently, as technological developments encouraged the consolidation of the industry, American companies quickly established a stranglehold. By 1898, the Boston Fruit Company and Minor Keith's companies accounted for 75 per cent of banana imports into the US (Read 1983: 191). Cashflow difficulties in the Keith group of companies saw it merge with Boston Fruit the following year to become the United Fruit Company (UFCo – renamed United Brands in 1968). UFCo's monopolisation of the US banana import trade was matched at the Central American level by territorial expansion, either through take-over, acquiring contracts in its own right from the governments of the day, or by taking equity-shareholdings in newly established ventures. By 1930, UFCo had established sole control over banana exports from Costa Rica, Guatemala and Panama, and accounted for 73 per cent of Honduran exports (Ellis 1983: 51). Its estates covered over 1.4 million hectares, although only 76,000 hectares were in operation at the time and it has traditionally exerted a tremendous influence over government policy.

The 1970s saw UFCo's attention move towards developing its newly established South American plantations. This was in response to scandals regarding bribery of Honduran government officials (Acker 1988: 66), hurricane destruction and renewed fungal outbreaks.

While the banana export trade had originated in Jamaica, the Caribbean economies had been increasingly squeezed out of the market by the Central American producers in the first half of the twentieth century. In the 1950s, however, smallholder banana cultivation began in earnest in the Windward Islands, growing to account for 49.4 per cent of the islands' export earnings by 1986 (Thomas 1989: 25), although the scarce and mountainous nature of land in the islands has meant that the industry is heavily dependent upon preferential access to UK and EU markets (Box 7.3).

Coffee

Coffee, originally sourced by the European powers exclusively from Muslim traders, was introduced into the region thanks to the activities of French (French Caribbean – 1715/6), British (Jamaica 1730) and Spanish (Cuba and Puerto Rico 1748–55) traders. With coffee plants quickly colonising the mountainous slopes of the Caribbean islands, Haiti provided 60 per cent of the world's coffee in 1769, the Caribbean providing the bulk of European coffee imports during the eighteenth century (Lundahl 1992: 111). As European demand grew, however, the relative scarcity of suitable terrain in the Caribbean saw the focus of production shift southwards to

Box 7.3   Bananas and the EU trade dispute

Preferential access to European markets is a commitment made to African, Caribbean and Pacific states (the ACP countries) under the Lomé Convention. In the instance of bananas, such arrangements accounted for 88 per cent of the UK market, with imports from Jamaica, the Windward Islands – St Lucia, St Vincent and Dominica – Belize and Suriname. The Windward Islands and Somalia provided 14 per cent of the bananas for the Italian market, and Cameroon and the Ivory Coast 35 per cent of bananas for the French market (Sutton 1996: 1–2). Imports from non-ACP countries were subject to quotas and/or a 20 per cent common external tariff (CET). In contrast, Germany, Denmark and Belgium sourced bananas entirely from Latin American growers, with the Netherlands and Ireland not far behind (90 per cent from Latin American suppliers).

Critically, however, smallholder banana production on rugged, mountainous slopes in the Caribbean is unable to compete with Central American (CA) plantation production on cost terms – Sutton (1996: 2, 3) estimating Caribbean production costs at around 0.555 ECU/kg in 1992 compared to the CA average of 0.2 ECU/kg in 1992. Consequently, while the Lomé agreement encouraged a number of ACP countries to expand production – to the extent that banana exports had grown to account for between 40 and 59 per cent (Dominica, Guadeloupe, and St Vincent) or 60 to 80 per cent (Martinique and St Lucia) of total exports in 1993/4 (Hallam 1997: 2) – it also resulted in higher costs, and lower consumption in the UK and the other quota-restricted markets (Borrell 1996: 3).

As moves quickened to establish a Single European Market by 31 December 1992, it became imperative to establish some kind of banana policy harmonisation – the question was how? While the removal of quotas and tariffs would force Caribbean bananas out of the European market – with all the attendant consequences for national economies and local producers – conversely quota/tariff retention penalised the low-cost Central and South American producers. A February 1993 compromise solution offered Latin American producers a quota of 2 million tonnes (subject to a tariff of 20 per cent) with any imports in excess of this being liable to a 170 per cent tariff. The problem seemed to have been resolved between the EU and the Latin American producer nations when the US, at the behest of the US fruit companies, became involved and petitioned the World Trade Organization (WTO) for redress. In May 1997, the WTO reported that the 1993 banana regime violated the principles of free trade on 26 counts and requested that the EU comply with WTO rules by 1 January 1999. The June 1998 EU response was rejected as purely 'cosmetic' by the US, which introduced retaliatory tariffs on a range of non-associated products, including Scottish cashmere sweaters and Parma ham (Barclay 1999: 12). The dispute now appears to be resolved, an amended quota system paving the way for the introduction of a flat tariff system in 2006.

Sources: Barclay (1999); Borrell (1996); Hallam (1997); Sutton (1996).

Brazil and Venezuela. In Central America, the lack of infrastructure was the biggest impediment to the development of the industry. The Caribbean coast was extremely inhospitable and housed only scattered populations, while the mountainous regions most suited to coffee cultivation were located in the interior. Nevertheless, the potential profits encouraged the emerging postcolonial elites to cultivate the crops. These coffee elites often emerged from immigrant (British, Spanish, German) backgrounds. However, such migrants, unlike their counterparts in the sugar and banana plantocracy, settled in the region and married into established local families, becoming naturalised Central American citizens over time (Paige 1997: 15). This pattern of settlement and marriage reflected the isolated locations of coffee farms. Guatemala (1853), El Salvador (1856) and Nicaragua (1860s) also started exporting coffee after the mainstay of each economy's colonial trade – cochineal and/or indigo – was destroyed by the development of synthetic substitutes. The exception was Honduras, where the almost complete absence of a transport infrastructure, labour scarcity and historic land tenure structures precluded the country from participating significantly in the coffee boom until the following century (Baumeister 1990: 41).

Bulmer-Thomas (1994: 18) argues that the expansion of coffee cultivation in the isthmus helped consolidate the stability of the political order as members of the elite sought to enhance their wealth and prestige by economic rather than militaristic means. As coffee revenues grew, so did the demand for commercial banking facilities, the Banco Anglo-Costarricense (Anglo-Costa Rican Bank) being the first to open its doors in the region in 1864. It also triggered railroad investment, with governments and private companies financing the construction of railroads intended to link the important inland coffee-growing regions with the Atlantic (Costa Rica) or Pacific (Nicaragua and El Salvador) ports. As the interests of elite and state became inextricably entwined, land and labour reforms and a barrage of fiscal incentives in the 1870s in Guatemala, El Salvador and Nicaragua not only provided a further stimulant to coffee expansion, but initiated a process of peasant expulsion (Brockett 1998: 19). By the end of the nineteenth century Central American coffee accounted for around 10 per cent of the global coffee trade, and had become the region's most important export (Williams 1994: 15–6).

Again, technological developments in the transport field were a key element behind the ascendancy of Central America as the main producer of coffee in the region. As steam replaced sail in the latter part of the nineteenth century, and bigger vessels reduced unit transport costs and ended the dependence upon favourable winds, trade with Central America became a more attractive prospect. The opening of the trans-Panamanian railroad in 1854 not only slashed transportation costs for the emerging Central American coffee trade, but provided a stimulus for national infrastructural investment – most notably the San José–Limón railroad in Costa Rica.

As the infrastructural deficit was addressed, the region was now able to exploit the vast tracts of mountainous land (500–2,000m) stretching from Northern Panama up to the southern state of Chiapas in Mexico and the deep volcanic soils upon which the coffee bush thrives. Such soils allow the production of premium *arabica* coffee – in contrast to the lower quality *robusta* beans produced in the lower-lying coffee-growing areas of Brazil and Africa. Coffee seedlings take around four to five years to mature, after which the bush will produce, disease permitting, for a further 15 to 20 years, although diminishing returns are evident in the latter years. Unlike sugar and bananas, which can be harvested throughout the year, coffee must be harvested shortly after the coffee cherries ripen (generally October to March). Processing necessitates removing the outer protective layers – either by laying out in the sun and then crushing to remove the inner parchment shell (relatively labour-intensive and likely to lead to a lower-quality product) or by 'washing' and threshing (more capital-intensive). Finally the beans are sorted and put in sacks ready for transfer to the roaster (*tostadora*) and/or to the exporting house (*exportadora*). Although significant economies of scale encouraged industrial concentration at the processing level, there are reduced pressures towards vertical integration due to the non-perishable nature of the product. It was the lack of pressure for vertical integration, allied to the high labour demand at harvesting time, that led to the long-lasting and markedly different tenure structures evident in the coffee sector across the region (see Table 7.3).

Although maintenance of established coffee farms requires relatively little in the way of labour inputs, start-up and harvesting requirements are highly labour-intensive. In Guatemala, El Salvador and, to a lesser extent, Nicaragua, the problem of labour supply was resolved in a twofold manner. First, legislation approved in the 1870s and 1880s abolished *ejidal* and communal landholdings, thereby freeing up lands for private coffee cultivation whilst depriving the local indigenous population of their traditional subsistence livelihood. However, this by itself was insufficient to guarantee the requisite labour supply, and so was reinforced with a panoply of labour laws that effectively forced workers on to the plantations. In Costa Rica, labour laws were less draconian and the emerging coffee elites were obliged to rely on salary inducements rather than coercion to source labour, a factor which militated against the emergence of a plantation sector of note. In Honduras (the maintenance of) and Mexico (the establishment of) *ejidal* systems of land tenure provided the majority of peasants with an opportunity to access land close to their community, thereby also frustrating the emergence of a significant coffee plantation sector in these two economies (Brockett 1998: 19–27; Harvey 1996: 192–6; Williams 1994: 105). Plantations were historically dominant in Haiti, but these were supplanted by peasant production after the plantation economy was effectively destroyed during the independence struggles at the beginning of the nineteenth century (Lundahl 1992: 79; Meyer 1989: 33). However, Haiti is now the sole Caribbean economy where coffee remains an important export earner, with coffee accounting for 50 to 80 per cent of the country's export revenues during the 1950s, falling to 22 to 42 per cent during the 1980s.

The dependence upon coffee (and, to a lesser extent, bananas) left Central America deeply exposed when the export-led growth model collapsed through the combined effects of massive overproduction in Brazil in 1926/7 and the Great Depression (1929–31). This shows the vulnerability of the region's economies to external shocks; a pattern apparent since its

Table 7.3 Farm area under coffee production for selected Central American countries, by farm size (various years)

Type of farm	Farm area under coffee¹	Costa Rica (1955)		Nicaragua (1957–58)		El Salvador (1940)		Guatemala (1966)	
		No. of farms	% of total area	No. of farms	% of total area	No. of farms	% of total area	No. of farms	% of total area
Family	0–9.9	20,824	47.8	7,821	24.5	9,768	18.9	25–30,000	11.6
Small employer	10–49.9	979	22.1	1,256	22.6	1,322	27.4	606	4.6
Estate	50–99.9	101	8.6	314	19.2	263	16.4	1,148	17.2
Integrated producer	100+	83	21.6	212	33.6	192	37.3	636	66.5
Total		21,987	100.1	9,603	99.9	11,545	100	27,390–32,390	99.9

Source: adapted from Paige (1997: 60).
Note
¹ Area given in *manzanas* (= approx. 1.7 hectares)

incorporation into global trading networks. Coffee prices slumped to around one-third of their 1920s peak, remaining subdued until after the outbreak of the Second World War a decade later. Growers responded in the short-term by forming producer groupings, which subsequently sanctioned the emergence of national dictators in return for protecting coffee interests from labour and social unrest (Bulmer-Thomas 1994: 60). More important to the medium-term recovery of the sector, however, was the deal cut with the US in 1940. As the Second World War effectively ruptured trading links with the European coffee market, the Central American economies successfully petitioned the US to introduce a quota system that allowed Central America to dispose of its exportable surplus at favourable prices through US markets. Almost overnight, an industry historically depend-ent upon the tastes of European consumers shifted its allegiance towards satisfying the American public, increasing dependence on the US market, which is also apparent in the industrial and service sectors (see Willis, this volume).

Unfortunately, longer-term prospects proved to be bleaker as high coffee prices in the immediate postwar period simply induced producer coun-tries to plant more seedlings. The predictable saturation of global markets that occurred as these coffee bushes reached maturity in the late 1950s persuaded the five Central American producers to join with Mexico, Brazil, Colombia, the US (as the largest global consumer) and others to sign the International Coffee Agreement (ICA) in 1959. Although the ICA system of export quotas appears to have favoured the Central American economies during the 1960s and 1970s (Bulmer-Thomas 1994: 153), consumer tastes were also changing in favour of *arabica* coffee. Consequently, in 1989 Central America, supported by Mexico and the US, demanded that the ICA-assigned quotas reflect such changing tastes by reducing quotas assigned to *robusta* producers in favour of *arabica* producers (Portillo 1993: 388/9). The ICA collapsed after *robusta* producers failed to give ground and, while a new accord was signed in 1992, this has not prevented prices from continuing on a downward trend. Falling coffee prices were an integ-ral, albeit largely unacknowledged, element behind the Zapatista uprising in Chiapas, Mexico's principal coffee-growing state, containing 26 per cent of Mexican coffee growers (Harvey 1996: 192). Current prospects for enhan-cing regional export revenues through increased coffee exports appear to be slim.

Cotton and beef: the second phase

The rapid post-war growth of the US economy offered the region an un-precedented opportunity to diversify its agro-export base through the incorporation of new tradeable commodities. Mexico largely forwent the opportunity, however, intent upon building up its industrial manufacturing base swiftly (Willis, this volume). The Caribbean economies failed to exploit this opportunity too, although here it was more a case of reluctance to shed ties with the old colonial powers, allied to a largely exhausted land frontier, rather than a decision to implement an alternative development model. Instead, the benefits of the second agro-export phase were largely confined to the Central American economies, beef and cotton exports providing a useful supplement to the exchange-revenues generated by the traditional agro-export crops (see Table 7.4). This diversity within the region again highlights the importance of recognising the ways in which supposed global trends are experienced differently at a national and local level.

Cattle

Cattle hold a special place in the Central American agricultural sector. Traditionally, cattle were allowed to roam free and could be walked to market, sharply reducing transportation costs. For the smallholder, cattle represented vital collateral; the cow produced milk for sale or consumption, could be sold should the need arise, eaten if times were hard, and produced calves which augmented their investment.

However, traditional Central American breeds were to prove inadequate in terms of beef production after the expansion in US demand for beefburgers radically changed cattle farming in Central America during the early 1950s. The native breeds that had been around since the colonial era were gradually confined to peasant farms as ranchers demanded new breeds.

The Central American climate, where cattle could be fed on grass throughout the year, gave it a comparative advantage in beef production. There was one hitch, however. An increased concern for food safety in the US rendered most of the production of the Central American slaughterhouses unfit for export. A development programme was needed to encourage not only herd growth, but also the setting up of suitable slaughtering facilities as cattle often had to be walked for many days to reach the slaughterhouse, and unhygienic conditions were common in slaughterhouses and markets (Williams 1986: 78).

The state played a key role in promoting the industry, encouraging the development of infrastructure such as roads, bridges and port facilities. The gradual reduction in the cost of refrigerated transport and containerisation revolutionised the way the product was transported, reducing the time and money involved in despatching beef from packing house to factories in the US. Improvements in animal husbandry also aided the region's cattle-breeding programme. Cheaper and more widely available medicines reduced ticks, worms and other parasitic diseases. Developments in artificial insemination greatly improved reproduction rates and advances in seed technology resulted in the introduction of better quality forage grasses (Williams 1986: 79).

The expansion in banana, sugar and coffee production, although heavily reliant upon markets, also fostered the development of internal commodity markets for the product. The internal market for beef was, however, different. Between 1957 and 1978 domestic consumption of beef actually dropped as good quality meat was exported. Land that had previously been given over to the production of staple grains was gradually turned over to pasture, with the subsequent loss of employment and food security. In Nicaragua, for example, only 1 per cent of total farmed area was cultivated with annual

Table 7.4 Traditional and second-wave non-traditional agro-exports from selected Central American countries 1954–1997 (US$ million)

Date	Costa Rica			El Salvador		Guatemala				Honduras				Nicaragua			
	CF	B	BF	CF	C	CF	B	C	BF	CF	B	C	BF	CF	B	C	BF
1954	35.1	35.8	n.a.	92.0	6.5	74.2	20.3	3.7	n.a.	14.0	29.3	0.1	n.a.	25.1	0.1	16.8	n.a.
1960	44.7	20.3	4.3	72.6	15.8	70.8	17.3	5.8	0.2	11.8	28.2	0.7	1.1	19.2	n.a.	14.7	3.1
1962	48.4	21.1	2.7	75.6	31.6	67.1	6.2	15.1	3.8	11.4	35.1	2.1	2.6	15.4	0.7	31.3	5.9
1972	77.8	82.8	28.2	130.1	37.1	106	17.2	40.0	17.7	26.7	81.9	0.6	15.9	33.0	3.4	62.8	38.2
1982	241.5	228.1	53	402.5	44.2	354	61.1	79.7	32.4	153.4	219	6.4	33.9	124	9.8	87.2	33.8
1992	160.6	485.3	25.6	149.7	1.8	248.3	111	13.6	16.2	107.8	286.5	n.a.	19.1	45.7	10	26.2	41
1997	419.2	588	0.2	514.8	0.1	589.5	151	n.a.	1.5	295.5	121.4	4.0	3.1	123.5	15.8	3.0	36.5

Sources: Bulmer-Thomas (1994); FAO data (1999).
Key: CF = coffee (green and roasted); B = bananas; C = cotton (lint); BF = beef (beef and veal)
Note: Cotton exports from Costa Rica and banana and beef exports from El Salvador were insignificant over the period

crops (down from 8 per cent in 1963), with pasture increasing from 39 per cent of farm area to 94 per cent over the same period (Williams 1986: 131). Credit was an integral part of the formula (Kaimovitz 1997: 57) suggesting that as a direct result of subsidised credit and direct loans to the sector, herd size increased from 4.7 million in 1950 to peak at 15.2 million in 1970. US beef imports from the region increased from about 15 million pounds in the late 1950s to a peak of about 260 million pounds by 1979.

Quotas were introduced to protect North American cattle farmers from a rapid rise in the availability of cheap beef imports. These favoured the large-scale investor as setting up a slaughterhouse involved a large capital outlay, and because political leverage was needed to get access to the US market as quota allocations went to established slaughtering companies. As a result, some of the largest investors in the region, such as Del Monte and Standard Fruit, became involved in the beef trade (Williams 1986: 104).

The expansion of cattle farming introduced three particular issues on to the Central American agrarian agenda. First, and most important, was deforestation. Kaimowitz (1997: 51) estimates that between 1950 and 1990 the area under forest in Central America fell from 29 million hectares to 17 million, largely as a consequence of herd growth. Second, cattle are a poor employer; compared to cotton, sugar and coffee, cattle ranching offers six, seven, and thirteen times less employment per hectare. Finally, due to a rapid decline in credit provision to the sector in the late 1970s, civil unrest in the region and the rapid rise in the cost of inputs due to rising oil prices, Central American beef has become less competitive on the global market (Kaimowitz 1997: 56).

Cotton

Cotton was not a total newcomer to Central America. It had been grown in the region in comparatively small quantities since the nineteenth century, the output being largely destined for the domestic garment industry. However, when the US civil war starved textile mills in the north of their raw material, El Salvador stepped in to bridge the gap; cotton represented 24 per cent of total Salvadorean exports in 1865, compared to just 4.85 per cent for coffee (Arias Peñate 1988: 201). After the American Civil War ended and US producers re-entered the world market, the price of cotton dropped dramatically, pricing Central American cotton out of the global market. However, there was a dramatic increase in cultivation in Central America following the Second World War as the region began to exploit a comparative production advantage: a large supply of cheap labour.

A number of technological advances helped boost cotton production along the fertile plains of the Pacific Ocean. First, new developments in agricultural technology resulted in cheaper and more reliable pesticides and fertilisers. Unfortunately, these technological benefits proved short-lived. As entomological resistance increased, so did the use of insecticides. Between 1964 and 1965 growers in Nicaragua were spending 40 per cent more on insecticides and 50 per cent more on fertiliser than they had been during the 1962–3 period. The manufacturers of fertilisers and insecticides, however, reaped the benefit as sales of fertilisers grew fourfold and sales of insecticide doubled (Williams 1986: 40). Second, road developments, as much a cause as a consequence of the cotton boom, greatly increased the amount of paved roads in the region, opening up previously inaccessible parts of the countryside (Williams 1986: 23). Economic incentives such as the provision of credit with preferential interest rates helped expand cotton cultivation. By the mid-1960s, 68 per cent of Central America's agricultural credit was devoted to cotton. In El Salvador and Honduras, cotton growers' co-operatives received government loans, while in Guatemala the private sector was more heavily involved following the overthrow of Guatemalan President Jacobo Arbenz in 1954. Cotton played a key role in modernising Central American agriculture in the postwar period: tractors, fertiliser and modern farming methods were not only used on cotton estates, but were switched into supporting the production of other crops.

Like other export crops in the region, the cotton industry was subject to high levels of vertical integration. Production, harvesting and processing, and export operations were often interlinked: partly because it helped reduce transaction costs, but also because it secured a regular supply for the large textile concerns in the US and Europe. El Salvador remained the one exception to the rule; here, a growers' co-operative established in the 1940s effectively controlled the buying and selling of cotton, removing power from the middlemen and merchants (Williams 1986: 38).

Cotton cultivation expanded rapidly. A regional output of 25,000 bales in 1940 rose progressively to 300,000 bales (1955), 600,000 bales (1962) before reaching 1.7 million bales in the late 1970s as the region became the third largest producer in the world. The profits to be made from cotton were so high that urban professionals were able to become part of the landed 'elites' through prodigious buying or renting of cotton lands. Even Guatemalan President Arbenz was credited with being one of the new urban cotton growers (Williams 1986: 34). Production expanded, land concentration and landlessness became ever more pronounced, while the fortunes made increased – until a combination of global events severely affected producers in the 1970s. First, prices have dropped steadily since their 1970 peak, thanks to a combination of expanded production in other parts of the world and an increased use of synthetic fibres. Second, imported inputs (farm equipment, seed, fertilizer and pesticides) were hit hard by the 1973 and 1979 oil price rises. Civil unrest was also seriously affecting production. In Nicaragua, exports fell from 113,600 metric tonnes in 1979 to just 19,644 the following year. Similarly Guatemala saw exports drop by one-third in 1986–87 at the height of the civil war, whilst continued civil unrest in El Salvador prevented cotton exports from reaching the levels recorded in the mid-1980s (see Willis and McIlwaine, this volume, for a discussion of these civil wars).

The agrarian reform 'backlash'

While much of the agricultural development outlined above led to the amassing of great wealth by national landed elites and foreign (particularly US) agro-food companies, another form of agricultural development, land redistribution, not only offers income-generating opportunities for the poorest in society, but helps to reduce political discontent in the countryside. Many of the region's countries, particularly on the mainland, have a history of land and agrarian reform, but the success and longevity of such policies vary across the region.

Although the Mexican revolution introduced a constitutional commitment to land redistribution (Box 7.2, page 163), the real redistributional impetus can be traced to the government of Lázaro Cárdenas (1934–40). Cárdenas redistributed around 20 million hectares to over 810,000 beneficiaries during his term, with the percentage of land in *ejidos* growing to represent 49 per cent of the arable land area, while the number of landless labourers halved (Thiesenhusen 1996: 37). Despite this, cattle haciendas continued to dominate the Mexican rural landscape.

The post-Second World War quest for industrialisation in Mexico (Willis, this volume) downplayed the importance of redistribution in favour of guaranteeing cheap food supplies for urban areas, focusing state support on high-productivity enclaves and the extension of the agricultural frontier through irrigation policies. These policies were initially successful as Mexican agricultural production increased 325 per cent between 1934 and 1965 (the largest increase in Latin America over the period), but this growth rate could not be sustained. The government announced a new food strategy in 1980 as food imports became commonplace (reaching 1 million tonnes a month in the 1970s), and growing landlessness (due to rural population increase, land degradation and the expansion of large-scale agricultural concerns) led to renewed bouts of land redistribution. The Mexican Food System (*El Sistema Alimentario Mexicano* – *SAM*) was designed to achieve self-sufficiency in maize and beans production within two years. Unfortunately, the austere economic policies introduced in the wake of the 1982 debt crisis (see Willis and McIlwaine, this volume) prevented the programme's continuation and any remaining pretence at agrarian reform was also shelved, to be replaced a decade later by a more market-driven policy approach.

Elsewhere in the region, pressures for land reform can be traced to the post-Second World War period when the introduction of new export products, particularly the land-extensive ranching of beef cattle, accelerated the process of peasant expulsion. Peasant populism was reinforced theoretically by the then dominant structuralist school of thought that argued that the regional agrarian structure was not only inegalitarian and inefficient, but exacerbated political and social tensions and provoked environmental degradation (Kay 1998: 5–6). Agrarian reform was viewed as a way of resolving these tensions, giving landless households a stake in society and a productive asset. If state support (credit, subsidised inputs, guaranteed marketing channels and prices) was extended to aid beneficiaries exploit this asset, output would grow, income distribution would improve and a wider domestic market would be created to the benefit of the local import substitution industrialisation sector (see Willis, this volume).

The first opportunity to test this thesis came with the election of Jacobo Arbenz in Guatemala. A manifesto commitment to agrarian reform was translated into reality with the approval of Decree 900 in 1952. By 1954 the programme had benefited a third of the country's peasantry and affected

Table 7.5 Impact of agrarian reform in Middle America

Country	Percentage of land affected	Percentage of the peasantry affected
Mexico	43	66
Guatemala (1952–54)	30	33
Cuba	83	71
Costa Rica	7	5
Honduras	12	9
El Salvador	19	13
Nicaragua	28	22

Source: Baumeister (1992: 21).

30 per cent of cultivated land (Table 7.5). However, it was vigorously opposed by the UFCo, which was offered a 'derisory' sum as compensation for the expropriation of 165,000 hectares, a reaction supported by the US State Department (Gleijeses 1992: 364–5). This helped to reinforce the US administration's view that Arbenz should go and, following Arbenz's overthrow in 1954, the reform was reversed, an estimated 62 per cent of the redistributed land being restored to its original owners (Goitia 1991: 169).

This momentarily closed the door to radical redistributionist programmes on the mainland. However, in Cuba, Castro's rise to power saw land reform used as a key instrument in transforming the rural economy. The original 1959 reform saw tenants, sharecroppers and squatters given title to the land they cultivated (Benjamin *et al.* 1984: 152). Private sugar plantations and ranches, which accounted for some 44 per cent of farmland, were confiscated and placed in a state farm/co-operative sector rather than redistributed, for fear that redistribution would adversely affect the country's export trade. A subsequent 1963 law saw the state expand its holdings to around 63 per cent of farmland, this rising to over 80 per cent by the end of the century as Cuban law prevents private land transfers, obliging prospective private vendors to sell to either the state farms or co-operatives (Azicri 1988: 163; Martinez-Alier 1977: 113).

The Cuban revolution offered renewed hope for agrarian reform in the region, US President J.F. Kennedy endorsing land reform as a legitimate weapon to be deployed to halt the spread of communism within the region. However, although the 1961 US Alliance for Progress programme made funds available for revising/introducing land tenure legislation, most governments elected to promote colonisation as opposed to redistribution schemes. There were some exceptions, notably Honduras and the Dominican Republic, where redistribution was introduced, while the government in El Salvador failed to promote either until the 1980s (Brockett 1998: 138; Goitia 1991: 170; Khan *et al.* 1996: 89; Ruben 1994: 154).

In Honduras, the 1962 Agrarian Reform Law introduced by the leftist-leaning government of Villeda Morales laid down clear guidelines as to which land was to be expropriated and redistributed, and to whom – although little effective redistribution actually took place until the late

1970s (Thorpe 2000). In the Dominican Republic, the assassination of the dictator, Rafael Trujillo, in 1961 paved the way for the expropriation of his estates. Yet, while further agrarian legislation in 1972 extended the expropriation remit to include *latifundios* and unfarmed lands (Meyer 1989: 49), redistribution was hampered by US military occupation (1965) and an internal coup (1983).

A much more far-reaching agrarian reform was undertaken in Nicaragua, however, following the revolutionary triumph of the Sandinistas in July 1979. Twenty-one per cent of national farmland (1.2 million hectares), which had been owned by the deposed dictator and his followers, were confiscated and incorporated into a state farm sector. Agro-industrial enterprises confiscated from the same families were absorbed into the state sector. Co-operatives were also assigned a central role in the new government's rural development programmes (Deere *et al.* 1985: 83; Groot 1994: 100). The reforms not only failed to assuage peasant demands for land, however, they were also largely overturned in the wake of the Sandinistas' electoral defeat in 1990 (Brockett 1998: 179).

In the Caribbean, reform attempts were singularly unsuccessful, apart from the cases of Cuba and the Dominican Republic. Although the 1945 Moyne Commission advocated land settlement schemes as a means to ward off growing social discontent, these schemes focused on encouraging new export crops (such as bananas in the Windward Islands) rather than a domestic-oriented peasant-based agriculture. Moreover, they were of limited scope: distributing uneconomic plots on marginal lands at a high capital cost. This neglect of domestic agriculture has ensured that, at the beginning of the twenty-first century, the islands remain heavily dependent upon imported foodstuffs (Thomas 1988: 126).

Also by the twenty-first century, the agrarian reform process in Central America and the Caribbean has largely expired. In its stead, a macroeconomic focus on stabilisation and structural adjustment has been reflected in the evolution of specific sectoral adjustment policies designed to restore the region's pre-eminence as an exporter of tropical foodstuffs. Agrarian 'modernisation' as opposed to agrarian reform was the new totem.

Neoliberalism, the 'new' agrarian policy and non-traditional agro-exports: the third phase

SAPs were applied in similar ways to the agricultural sector as they were to other parts of the region's economies in the face of mounting debt and international pressure (see Willis and McIlwaine, and Willis, this volume). The main elements of the reform packages were price liberalisation, institutional reform, land titling as opposed to land redistribution, and an expansion of exports through diversification into non-traditional products.

Liberalising prices (both internationally and domestically) is seen as a key component within the new agrarian framework. It is argued that the global market can provide a strong monetary stimulus for producing internationally traded crops (tradeables). Consequently, neoliberal policymakers advocate moving from the over-valued (fixed) exchange-rates of the past

– which were seen as an essential part of the import substitution industrialisation strategy – to a depreciated, more 'realistic' (floating) exchange-rate. By restoring the incentive to export, such a policy is perceived to be of immense benefit to a region that has historically been heavily reliant upon agro-export trade (see Table 7.1 on page 160).

However, a number of problems have been identified with this approach. Firstly, export growth may be impeded by quotas (for example, coffee under the International Coffee Agreement, and Central American and Caribbean banana and sugar exports to the EU and US respectively). Secondly, growth may be limited because demand elasticities for primary agricultural exports are low on the international market. In addition, tariff removal may encourage food imports to the detriment of local production. This was a particular point of contention in the arrangements regarding agricultural products under NAFTA (Heredia and Purcell 1994: 17), the ensuing compromise being a 15-year transitional period for regional trade in maize and beans. Climatic conditions can also affect the success of agriculture under this liberalised price regime, and international price volatility may also create problems, overriding the expected benefits of currency realignment. Finally, although often ignored, is the question of who benefits from increasing the proportion of (i) land sown with export crops, (ii) crops diverted from the domestic to the international market (Thorpe 1997: 20, 1995: 99).

Neoliberal policymakers also reject domestic price controls on the grounds that keeping prices artificially low frustrates production of basic foodgrains for the domestic market. Instead, an elaborate price band system is proposed, which links movements in domestic foodgrain prices to shifts in international foodgrain prices. Finally, price liberalisation is extended to the input sector, most particularly agrarian credit provision. Subsidised rates of interest are withdrawn, preferential financing facilities are closed (although sometimes an exception is made for new non-traditional agro-export crops – NTAX) and credit provision reverts to being rationed by price. In Costa Rica, preferential interest rates for small and medium producers were seen as distortionary and eliminated early on during the 1980s adjustment programme, rural credit supply halving between the 1970s and 1980s as a consequence (Fallas Venegas and Rivera Urrutía 1988: 40; Thorpe 1997: 22). In Nicaragua, the 1990 sectoral adjustment programme not only saw nominal interest rates become the highest in the region, but also allowed large farmers to increase their share of agricultural credit from 31 per cent in 1990 to 71 per cent just 2 years later (Jonakin 1995: 12). Consequently, concern that neoliberal credit policy has squeezed/is squeezing out the poor has led a number of countries (Honduras and Mexico among them) to retain cheap credit lines for the poor.

Price liberalisation has been accompanied by the reform of state agricultural support institutions through the cutting back of expenditure on such organisations (Calva 1991: 108), and strategies to improve institutional efficiency. In the case of market-intervention institutions, the logical corollary of abandoning procurement and producer support schemes has been the downsizing or abolition of such enterprises. In Mexico CONASUPO (*Compañía Nacional de Subsistencias Populares* – National Subsidised Staple Products Company) first halved its grain purchases and then, following the deregulation and liberalisation of grain import procedures, completely withdrew from commercialisation activities (Luiselli 1988: 32; Matus and Vega 1992: 322). A similar tale has unfolded in both El Salvador and Nicaragua (Spoor 1997: 33), despite doubts being expressed about whether the state's withdrawal from the commercialisation sphere is likely to be filled by an effective network of private agents (Thorpe 1997: 24).

In the case of the public sector credit institutions, Clemens (1993: 26) has documented how, as state support was slashed, the volume of funds lent to Nicaraguan maize and bean farmers in 1992/3 fell to just 14 per cent of the 1991/2 level. Agrarian reform institutions were not exempted. Gates (1996: 46) delineates how SARH (*Secretaría de Agricultura y Recursos Hidráulicos* – the Secretariat of Agriculture and Water Resources), the lead agency in the Mexican *ejidal* sector, was not only rechristened in 1995 (as SEMARNAP, *Secretaría de Medio Ambiente, Recursos Naturales y Pesca* – the Secretariat of the Environment, Natural Resources and Fish) but suffered extensive staff cutbacks in the 1980s and 1990s as the neoliberal revolution gathered pace. Finally, privatisation has been a favoured mode of reducing state influence in the agro-processing and extension sphere (CENPAP 1993: 77; Thorpe 2000).

The third key area in which neoliberal policymakers have wrought substantive change is in land tenure policy. As outlined above, the commitment to direct expropriation and redistributive reform has been dropped in favour of a market-driven solution underpinned by individual title adjudication programmes. Titling programmes do not only afford security of tenure through clarifying the legal rights to own and dispose of land, but the resulting title granted can be deployed as collateral in order to raise working capital. The principal problem facing neoliberal strategists on the tenure front, however, related to past agrarian legislation that invariably promoted collective as opposed to individual title, and prohibited the sale or rental of reform sector lands. The most celebrated example of this was in Mexico, where agrarian reform was enshrined in Article 27 of the 1917 Constitution (Box 7.2). This failed to deter neoliberal reformers and, in early 1992, the Mexican Congress approved amendments to Article 27 to allow *ejidal* land to be titled and privatised, and joint ventures between *ejidatorios* became legal (Green 1996: 267). Although progress in providing individual title to *ejidatorios* has been slow, this did not deter several major Mexican agribusinesses (Maseca, El Trasgo and Gamesa) from signing joint-venture contracts with *ejidos* shortly after the amendments became law (Gates 1996: 51).

A similar reform in Honduras had more pronounced effects; within three years of the approval of the 1992 Agrarian Law, 254 co-operatives comprising 5,372 members had disposed of over 30,847 hectares of land (8.5 per cent of all co-operative land). The principal beneficiaries were the US fruit companies and domestic agro-industrialists geared towards the export sector as opposed to the landless or land poor (Thorpe 2000). Indeed, although an avowed aim of neoliberal tenure reform is to widen access to land by instituting land-banks along the lines of the Centavo Foundation in Guatemala, nearly a decade on, the commitment remains still-born in Honduras.

The final major element of SAP-related reform in the agricultural sector was increasing support for the cultivation of non-traditional agro-export (NTAX) crops. NTAX are agricultural commodities not previously exported – either because they have been newly introduced to the country, or because they were previously solely oriented towards satisfying domestic consumption requirements (Barham *et al.* 1992: 43). These NTAX include fruit (melon, mangoes, papayas, pineapples, nuts) and flowers and vegetables (broccoli, snowpeas, Brussels sprouts, cauliflowers among others). They are seen as complementing the portfolio of traditional agro-exports, consolidating the process of integration into global, but mainly US, markets. On one hand, they offer higher per-hectare returns for growers; Thrupp (1995: 60), for example, estimated that a 100-hectare farm cultivating flowers was capable of generating greater revenues than a 20,000-hectare farm sown with either cotton or sugar cane, while TED (1999) notes that net per-hectare returns on snowpeas were, on average, 15 times that of corn. On the other hand, supporters assert that they provide governments with much-needed export earnings and contribute to poverty reduction through rural employment generation.

The development and growth of the NTAX industry can be traced to the efforts of the US Agency for International Development (USAID), the World Bank and the Inter-American Development Bank (among others) which, with the support of national governments, established incentives and export promotion centres in order to reduce the production and marketing risk associated with such new ventures. International trade agreements such as the 1983 Caribbean Basin Initiative (see Willis and McIlwaine, this volume) and the Generalised System of Preferences, complemented domestic efforts by offering privileged access for named commodities to the US market. The most successful initial ventures were in Guatemala where, thanks to a series of USAID loans stretching back to 1970, the share of NTAX rose from under 6 per cent in 1980 to just over 22 per cent a decade later (Carletto *et al.* 1996: 1). Although Costa Rica also became a substantive producer of NTAX during the 1980s, NTAX growth in the rest of the region is more a 1990s phenomenon (Figure 7.1). A more specialised offshoot of this general trend in NTAX growth has been the recent attention paid to organic and fair trade produce (Box 7.4).

Trejos (1992: 96), however, is sceptical of the true worth of such NTAX programmes, arguing that (i) incentives are/have been given in an indiscriminate manner without regard to the size of the potential product market, (ii) marketing and technical economies of scale have been sacrificed through the pursuit of national, as opposed to regional, output goals, (iii) success is critically dependent upon linking with transnational companies who can resolve the problems of transportation and international marketing, and (iv) growth potential is constrained, in the short to medium term, by the lack of adequate infrastructural facilities (airport storage facilities, quality-control mechanisms, etc.).

Besides these more general considerations, it is important to point out that doubts also exist as to both the equity and environmental implications of cultivating NTAX (see also Barton, this volume). As the production of Guatemala's leading NTAX, snowpeas, increased more than sevenfold

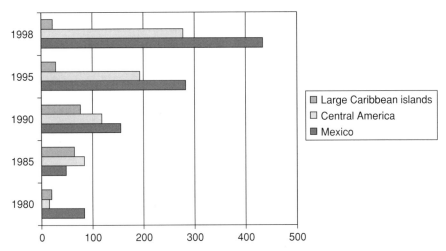

Figure 7.1 Non-traditional exports from Mexico, Central America and the Caribbean 1980–1998 (US$ million)
Source: FAO data, various years.
Large Caribbean islands: Jamaica, Haiti, Cuba, Dominican Republic, Trinidad and Tobago.
Small Caribbean islands are not shown as total NTAX exports over the period were comparatively insignificant (under US$100,000).
NTAX: green beans, nuts, broad beans, cabbages, melons, cauliflowers, citrus fruits, mangoes, spices, papayas, pineapples, strawberries.

between 1983 and 1991 (TED 1999: 2), so did pesticide usage. The absence of a real winter, the multitude of tropical pests and high US product quality standards quickly pushed producers on to a 'pesticide treadmill'. Pesticide costs per hectare even rose beyond those of the cotton plantations. Not only has such wide (and often indiscriminate) usage endangered workers' health and contaminated water-tables, it also led to the US rejecting 3,081 Guatemalan NTAX shipments (mostly of snowpeas) between 1990 and 1994 due to the application of chlorothalanil, an unregistered pesticide. Although the chance of shipment rejection has now been reduced with pre-shipment screening in Guatemala, this has imposed substantial costs on local producers (Thrupp 1995). The 'pesticide treadmill' is one of the reasons for this, despite the fact that 'while initially accessible to all farmers, only the better endowed households with more land owned and better quality land were able to persist in growing NTAXs' (Carletto *et al.* 1996: 20; see also Rosset 1991, 1992; TED 1999). Resource-rich households have therefore found themselves at an advantage in producing the input-intensive, knowledge-intensive NTAX and the small-scale resource-poor farmer is gradually being squeezed out of NTAX production, with all the attendant implications for rural equity. This exemplifies the validity of Sen's notions of 'entitlements' and how macro-economic changes are differentially experienced by various sectors of society, leading to greater poverty for some, but wealth creation for others (see McIlwaine, this volume).

Box 7.4 Organic agriculture and 'fair trade' exports

Although it is still in its infancy, the past decade has seen a rapid increase in the scale and extent of organic farming throughout the region, with growth rates estimated at somewhere between 25 and 30 per cent (Soil Association 1998). The underlying reasons are: (i) the rise in the cost of imported fertilisers and pesticides following the neoliberal-ordained currency devaluations, which has encouraged a shift towards low-input, organic methods (particularly in Cuba, as Rosset and Benjamin (1994) have detailed), and (ii) increased demand in the markets of North America and Europe for organically produced items. A third contributing factor is the growing market for 'fair-trade' products in the developed world: organics produced for an external market are often linked to fair-trade initiatives where producers are paid a fairer price for their produce. These prices are often much higher than the going market rate, hence a greater part of the profits go to the farmer, not the middlemen or retailers.

There is a diverse range of products grown organically in the region and traded in global markets: bananas, tropical fruit, nuts and cotton are all exported, with coffee and cacao heading the list. One of the most successful stories can be found in Belize, where Green and Black's Organic Chocolate became involved in cocoa production in 1993. Sales of organically produced cacao rose from 8,200 kilos in 1993 to 23,636 kilos by 1997, with income for the Toledo Cacao Growers Association – the principal participant in the venture – rising by 40 per cent since production began (Soil Association 1998). Green and Black's operate on a fair-trade basis and have further contributed to local sustainable production by paying farmers to plant tropical hardwood trees as shade for the cacao trees (Sams 1998). A more extreme example can be found in Grenada where Sainsbury's has unveiled plans to buy up all the organic tropical produce on the island in order to satisfy growing customer demand in the UK (*The Guardian* 2000).

Sources: *The Guardian* (2000); Rosset and Benjamin (1994); Sams (1998); Soil Association (1998).

Conclusion

In this chapter we have outlined the processes through which the region's nations and rural communities have been incorporated into global agricultural trading networks. The importance of both colonialism and the reliance on the US for both production organisation and demand is seen throughout the region, as is the impact of technological developments. However, it is clear that experiences of agricultural developments have not been homogeneous. There are intraregional differences in crops produced, the organisation of production and the success of land reform policies.

However, in macro-economic terms, almost all the region's nations are currently experiencing the application of neoliberal reforms to the agricultural sector. In a sense, this neoliberal 'market panacea' is simply the latest policy manifestation designed to support the ongoing process of agrarian globalisation in the region, seeking to put a new gloss upon the tarnished images of the 'banana republics' and 'sugar dependencies' of the past. Like its predecessors, it too is primarily concerned with encouraging the expansion of production (and specifically exports) rather than tackling the issues of rural inequality and poverty. Yet rural poverty and the related problem of landlessness is a very real concern in the region, sparking the Zapatista uprising in Mexico and prompting the formation of ASOCODE, a regional umbrella grouping through which Central American peasant unions are stridently pressing their demands for land. Thus while some agricultural sectors may be integrated into the global economy, there is another side to the coin: a rural subsistence agriculture, disarticulated from the market and largely impervious to globalisation processes, within which the majority of the poor are to be found (Thorpe 2000). Consequently, the market-oriented policies of the neoliberals are likely to have a minimal impact upon this grouping – and other more specific policies will be essential if serious inroads are to be made in reducing regional poverty levels.

Useful websites

www.fao.org/ag Home page for Agriculture 21 at the Food and Agriculture Organisation. Many links to a variety of topics on agriculture and food production.

http://apps.fao.org An invaluable database of statistics on food production, including harvests, exports, imports and labour. Works well and covers the entire world.

www.warwick.ac.uk/fac/soc/CSGR Home page of the Centre for the Study of Globalisation and Regionalisation. Useful downloadable papers on theoretical issues.

http://pw1.netcom.com/~hhenke/index.htm Home page of the Virtual Institute of Caribbean Studies.

http://www.worldbank.org/rural World Bank rural development topic page. The link to the 'Land policy' subsection provides an excellent unbiased overview of land policy.

Further reading

Brockett, C.D. (1998) *Land, Power and Poverty: Agrarian Transformation and Political Conflict in Central America*. Westview Press, Boulder, CO.
An excellent book linking discussion of agrarian reform, land tenure and politics in the region.

Paige, J.M. (1997) *Coffee and Power: Revolution and the Rise of Democracy in Central America*. Harvard University Press, Cambridge, MA.

A well-balanced and thorough examination of the development of coffee from a political-economy perspective.

Richardson, B.C. (1997) *Economy and Environment in the Caribbean: Barbados and the Windwards in the Late 1880s.* University of West Indies Press, Barbados, Jamaica and Trinidad and Tobago.
A useful overview of the economic development of parts of the Caribbean in the late nineteenth century.

Thorpe, A. (1995) Structural adjustment and the agrarian sector in Latin America, in Spoor, M. (ed.) *The 'Market Panacea': Agrarian Transformation in Developing Countries and Former Socialist Economies.* Intermediate Technology Publications, London: 15–28.
This chapter provides a good, balanced and unbiased overview of the economic policy background to the agrarian sector in the region.

References

Acker, A. (1988) *Honduras: the Making of a Banana Republic.* Between the Lines, Ontario.

Arias Peñate, S. (1988) *Los Subsistemas de Agroexportación en El Salvador. El Café, el Algodón y el Azúcar.* UCA Editores, San Salvador.

Ayala, C.J. (1995) Social and economic aspects of sugar production in Cuba, 1880–1930, *Latin American Research Review* **30**: 95–124.

Azicri, M. (1988) *Cuba: Politics, Economics and Society.* Pinter Publishers, London and New York.

Barclay, C. (1999) *The Trade Dispute between the EU and the USA over Bananas.* Science and Environment Section, House of Commons Library, Research Paper No. 99/28.

Barham, B., Clark, M., Katz, E. and **Schurman, R.** (1992) Non-traditional agricultural exports in Latin America, *Latin American Research Review* **28**(2): 43–82.

Baumeister, E. (1990) Problemas nacionales: el café en Honduras, *Revista Centroamericana de Economía* **11**(33): 33–78.

Benjamin, M., Collins, J. and **Scott, M.** (1984) *No Free Lunch: Food and Revolution in Cuba Today.* Institute for Food and Development Policy, San Francisco.

Borrell, B. (1996) *Beyond EU Bananarama 1993: the Story gets Worse.* Centre for International Economics, Canberra and Sydney.

Brockett, C.D. (1998) *Land, Power, and Poverty: Agrarian Transformation and Political Conflict in Central America.* Westview Press, Boulder, CO.

Bulmer-Thomas, V. (1994) *The Political Economy of Central America since 1920.* Cambridge University Press, Cambridge.

Busch, L. and **Juska, A.** (1997) Beyond political economy: actor networks and the globalization of agriculture, *Review of International Political Economy* **4**(4): 688–708.

Calva, J.L. (1991) The agrarian disaster in Mexico, 1982–9, in Twomey, M.J. and Helwege, A. (eds) *Modernisation and Stagnation: Latin American Agriculture in the 1990s.* Greenwood Press, Westport, CT: 101–120.

Carletto, C., de Janvry, A. and **Sadoulet, E.** (1996) *Knowledge, Toxicity, and External Shocks: the Determinants of Adoption and Abandonment of Non-Traditional Export Crops by Smallholders in Guatemala.* Dept. of Agricultural and Resource Economics, Division of Agriculture and Natural Resources, University of California, Berkeley.

CENPAP (1993) Aportes para la estrategia de desarrollo agropecuario en Nicaragua, in *Por la Busqueda de una Estrategia de Desarrollo para Nicaragua.* Escuela de Economía Agrícola-UNAN, Managua.

Clemens, H. (1993) La estrategia de desarrollo agropecuario en Nicaragua: Una visión desde la Universidad, in *Por la Busqueda de una Estrategia de Desarrollo para Nicaragua.* Escuela de Economía Agrícola-UNAN, Managua: 11–30.

Deere, C., Marchetti, P. and **Reinhardt, N.** (1985) The peasantry and the development of Sandinista agrarian policy, 1979–84, *Latin American Research Review* **20**(3): 75–109.

ECLAC (1999) *Anuario Estadistico de Latino America y el Caribe 1999: Part I Indicators of Economic and Social Development in Labin America and the Caribbean.* ECLAC, Santiago.

Ellis, F. (1983) *Las Transnacionales del Banano en Centroamérica.* Educa, San José.

Fallas Venegas, H. and **Rivera Urrutía, E.** (1988) *Agricultura y Cambio Estructural en Centroamérica.* IICA Series Documentos de Programas No. 8, San José.

FAO (2000) *On-line Agricultural Statistics databases (http://apps.fao.org/page/collections?subset=agriculture).*

Gates, M. (1996) The debt crisis and economic restructuring: prospects for Mexican agriculture, in Otero, G. (ed.) *Neo-liberalism Revisited: Economic Restructuring and Mexico's Political Future.* Westview Press, Boulder, CO: 43–62.

Gleijeses, P. (1992) La reforma agraria de Arbenz, in Cambranes, J.C. (ed.) *500 Años de Lucha por la Tierra, Vol.1.* FLACSO, Guatemala: 349–378.

Goitia, A. (1991) Reforma agraria con orientación de mercado, in Ruben, R. and Van Oord, G. (eds) *Más Allá del Ajuste: La Contribución Europea al Desarrollo Democrático y Duradero de las Economías Centroamericanas.* DEI, San José: 167–194.

Green, L. (1996) What's at stake? The reform of agrarian reform in Mexico, in Randall, L. (ed.) *Reforming Mexico's Agrarian Reform*. M.E. Sharpe, New York: 267–270.

Groot, J.P. de (1994) Reforma agraria en Nicaragua: Una actualización, in Groot, J.P. de and Spoor, M. (eds) *Ajuste Estructural y Economía Campesina: Nicaragua–El Salvador–Centroamerica*. ESECA-UNAN, Managua: 97–122.

The Guardian (2000) Supermarket Isle, Wednesday, 19 January 2000.

Hallam, D. (1997) *The Political Economy of Europe's Banana Trade*. University of Reading, Dept. of Agricultural and Food Economics, Occasional Paper No. 5.

Harvey, N. (1996) Rural reforms and Zapatista rebellion: Chiapas, 1988–1995, in Otero, G. (ed.) *Neo-liberalism Revisited: Economic Restructuring and Mexico's Political Future*. Westview Press, Boulder, CO: 187–208.

Heredia, C.A. and **Purcell, M.E.** (1994) *La Polarización de la Sociedad Mexicana: Una Visión desde la base de las Politicas de Ajuste Económico del Banco Mundial*. Grupo de Trabajo de las ONGs sobre el Banco Mundial, Mexico. D.F.

Jonakin, J. (1995) Agrarian policy and crisis in Nicaragua's political transition. Paper presented at the XIX International LASA Conference, Washington, D.C., 28–30 September.

Kaimowitz, D. (1997) Policies affecting deforestation for cattle in Central America, in de Groot, J.P. and Ruben, R. (eds) *Sustainable Agriculture in Central America*. Macmillan, Basingstoke: 51–60.

Kay, C. (1998) *The Complex Legacy of Latin America's Agrarian Reform*. Institute of Social Studies, Working Paper Series No. 268, The Hague.

Khan, A.A., Carías, M.V. and **López, M.A.** (1996) Reforma agraria y el acceso a la tierra, in Khan, A.A. (ed.) *Economía Agrícola: Recorrido Teorico y Debates de Interes Actual*. Prografic, Tegucigalpa: 57–102.

Knight, F.W. (1978) *The Caribbean: the Genesis of a Fragmented Nationalism*. Oxford University Press, New York.

Luiselli, C. (1988) *Las Políticas de Ajuste Estructural sobre el Sector Agro-alimentario en México, Ajuste Macroeconómico y Sector Agropecuario en América Latina*. IICA, Buenos Aires.

Lundahl, M. (1992) *Politics or Markets? Essays on Haitian Underdevelopment*. Routledge, London.

Markiewicz, D. (1993) *The Mexican Revolution and the Limits of Agrarian Reform, 1915–1946*. Lynne Rienner, Boulder, CO.

Martinez-Alier, J. (1977) *Haciendas, Plantations and Collective Farms: Agrarian Class Societies: Cuba and Peru*. Frank Cass, London.

Matus, J.G. and **Vega, D.** (1992) Política macroeconómica y sectorial, sus reformas y la pobreza rural en México, in Trejos, R. (ed.) *Ajuste Macroeconómico y Pobreza Rural en América Latina*. IICA, San José: 309–352.

Meyer, C.A. (1989) *Land Reform in Latin America: the Dominican Case.* Praeger, New York.

Mintz, S. (1986) *Sweetness and Power: the Place of Sugar in Modern History.* Harmondsworth, Penguin, New York.

Paige, J.M. (1997) *Coffee and Power: Revolution and the Rise of Democracy in Central America.* Harvard University Press, Cambridge, MA.

Portillo, L. (1993) El convenio internacional del café y la crisis del mercado, *Comercio Exterior,* Abril, 378–91.

Posas, M. and **Del-Cid, J.R.** (1983) *La Construcción del Sector Público y del Estado Nacional en Honduras, 1876–1979.* Educa, San José.

Randall, S.J. and **Mount, G.S.** (1998) *The Caribbean Basin and International History.* Routledge, London.

Read, R. (1983) The growth and structure of multinationals in the banana export trade, in Casson, M. (ed.) *The Growth of International Business.* Allen and Unwin, London: 180–213.

Richardson, B.C. (1997) *Economy and Environment in the Caribbean, Barbados and the Windwards in the late 1880s.* The University of the West Indies Press, Barbados, Jamaica and Trinidad and Tobago.

Rosset, P.M. (1991) Sustainability, economies of scale and social instability: Achilles heel of non-traditional export agriculture? *Agriculture and Human Values* **8**(4): 30–37.

—— (1992). 'Economies of Scale and Small Producers: the Case of Non-Traditional Export Agriculture and Structural Adjustment in Central America'. Paper presented at XVII International LASA Conference, Los Angeles, California, 24–27 September.

Rosset, P.M. and **Benjamin, M.** (eds) (1994) *The Greening of the Revolution: Cuba's Experiment with Organic Agriculture.* Ocean Press, Melbourne.

Ruben, R. (1994) Reforma agraria y transformación del campesinado en El Salvador: perspectivas para la consolidación empresarial del sector reformado, in Groot, J.P. de and Spoor, M. (eds) *Ajuste Estructural y Economía Campesina: Nicaragua–El Salvador–Centroamerica,* ESECA-UNAN, Managua: 149–178.

Sams, C. (1998) Sustainable chocolate: a practical business example, *ILEIA Newsletter,* 12 December.

Soil Association (1998) *Organic agriculture and sustainable rural livelihoods in developing countries.* Report prepared for DFID by the Soil Association.

Spoor, M. (1997) Agrarian transformation in Nicaragua: market liberalisation and peasant rationality, in Spoor, M. (ed.) *The 'Market Panacea': Agrarian Transformation in Developing Countries and Former Socialist Economies.* Intermediate Technology Publications, London: 29–42.

Sutton, P. (1996) 'In Whose Interest? The Banana Regime of the European Community, the Caribbean and Latin America'. Paper prepared for the

. Annual Conference of the Society of Latin American Studies, University of Leeds, 29–31 March.

Tannenbaum, F. (1968) *The Mexican Agrarian Revolution*. Archon Books, Washington, D.C.

TED (1999) Non-traditional agricultural exports and the pesticide problem in Guatemala (available at *gurukul.american.edu/TED/SNOWPEA.HTML*, accessed 6 July 1999).

Thiesenhusen, W.C. (1996) Mexican land reform, 1934–91: Success or failure?, in Randall, L. (ed.) *Reforming Mexico's Agrarian Reform*. Columbia University Seminar Series, M.E. Sharpe, Armonk, New York: 35–47.

Thomas, C.Y. (1988) *The Poor and the Powerless: Economic Policy and Change in the Caribbean*. Latin America Bureau, London.

Thomas, M. (1989) *Cash Crops and Development: Bananas in the Windward Islands*. IDS Discussion paper No. 258, University of Sussex.

Thorpe, A. (1997) Structural adjustment and the agrarian sector in Latin America, in Spoor, M. (ed.) *The 'Market Panacea': Agrarian Transformation in Developing Countries and Former Socialist Economies*. Intermediate Technology Publications, London: 15–28.

—— (2000) *Agrarian Modernisation in Honduras*. Edward Mellen Press, Lampeter.

Thrupp, L.A. (1995) *Bittersweet Harvests for Global Supermarkets: Challenges in Latin America's Agricultural Export Boom*. World Resources Institute, Washington, D.C.

Trejos, R. (ed.) (1992) *Ajuste Macroeconómico y Pobreza Rural en América Latina*. IICA, San José.

Watts, M. and **Goodman, D.** (1997) Agrarian questions: global appetite, local metabolism: nature, culture and industry in *fin-de-siècle* agro-food systems, in Goodman, D. and Watts, M. (eds) *Globalising Food: Agrarian Questions and Global Restructuring*. Routledge, London: 1–32.

Whatmore, S. (1995) From farming to agribusiness: the global agro-food system, in Johnston, R.J., Taylor, P. and Watts, M. (eds) *Geographies of Global Change*. Blackwell, Oxford: 36–49.

Williams, R.G. (1986) *Export Agriculture and the Crisis in Central America*. University of North Carolina Press, Chapel Hill.

—— (1994) *States and Social Evolution: Coffee and the Rise of National Governments in Central America*. University of North Carolina Press, Chapel Hill.

CHAPTER 8

Unsustainable development: environmental protection issues in rural and coastal areas

Jonathan R. Barton

Introduction: sustainable development or environmental crisis?

This chapter explores some important dimensions of environmental protection in Middle America. Underlying the discussion are three key trends that are fundamental in understanding environmental issues in Middle America. The first relates to globalisation and has a series of contradictory outcomes. On the one hand, it will increase pressure on the environments of Middle America for agricultural exports, and on the other it will offer new opportunities with consumer preferences for environmentally friendly products and services, for example organic fruit and vegetables, and low-impact tourism. A second crucial process is sustainable development with its guiding principle of inter-generational equity (particularly in terms of environmental resources) and its recognition of the symbiosis between socio-economic development and environmental change. Environmental protection and conservation is a third process that plays a key role in discussions of both globalisation and sustainable development. Since the region is so dependent on its natural resource base for production and services, the environment has to be protected and managed alongside longer-term objectives of social development and equity, and the promotion of alternative economic activities.

All these trends are heavily influenced by economic and social change and are related to environmental change since most livelihoods in the region remain tied to natural resource exploitation. These include traditional activities such as small-scale farming and artisanal fishing, and more modern ones such as the monoculture model of agricultural and livestock production, industrial forestry practices, the shift towards non-traditional products for export markets, and the demands from increasing service exports via tourism. The management of these activities will determine whether the region will move from its current path of unsustainable development, characterised by extensive environmental degradation and high levels of poverty, to one of sustainability via environmental protection and a less expropriative relationship with the global economy.

The extent to which the region's environments are in crisis or not is the basis of a complex debate. Any explanation lies within the wider political issues about the motives behind environmental exploitation and the relationship between environmental degradation, economic development and poverty (see Annis 1992; Faber 1993), as well as the influence of natural disasters. There are several key elements to the question, such as: whether existing activities have exceeded the carrying capacity of the environments in which they are operating (in other words, whether the natural environment is able to sustain its ecological diversity and equilibrium in the face of these activities); the ability of the environments to recover from current levels of degradation through their absorptive and adaptive capacities; and finally, how crises can be mitigated or transformed in order to promote sustainable practices in terms of ecology, human health and livelihoods.

Rather than present the entire region as experiencing an 'environmental crisis', it is more productive to categorise different environments, to recognise the diversity that exists, to stress that levels of degradation are highly variable, and to note that responses should be specific rather than generic. Middle America has many environments at different geographical scales of analysis, each of which is experiencing different human and natural pressures. Some environmental issues are regionwide, such as natural disaster vulnerability; others are particularly national, such as Haiti's soil erosion crisis or Costa Rica's trade-off between cattle ranching and national parks; and in turn, others are more local in origin, focused on catchments, topography, soil type, intensity of human impact, etc.

It is true to say that some of these environments are indeed 'in crisis', in that they are exploited beyond their natural capacity for ecological regeneration. In these circumstances, the quality of these environments is deteriorating in terms of stability, biodiversity and utility. Fragile tropical forest environments in the region are often cited as examples of this condition. Other environments are 'under threat', in that current levels of exploitation are having degrading impacts but within the limits of regeneration at present. Particular areas under cultivation provide suitable examples since changing agricultural practices or techniques – perhaps to slow erosion rates or restore soil fertility – could reduce the threat to these environments. For instance, Cuba has introduced techniques such as contour ploughing and cover cropping as part of its more agro-ecological approach to food production. Also, Loma Linda, a Honduran non-governmental organisation (NGO), has promoted a no-till system whereby weeds are left in the soil, which allows for good crop yields with no need for chemical inputs and no significant soil loss (Altieri 1995; Rosset 1996). Naturally there are still environments that exist in a state of 'ecological equilibrium' and would not be considered as 'under threat' at present. However, the concern is that, without protection, current rates of exploitation and the opening-up of forest frontiers for agriculture, ranching and logging will threaten these environments in the near future.

A more positive note in terms of typology relates to those environments that are being protected, conserved, remediated or restored. These may be termed 'ecologically managed' environments. The initiatives in the region to establish national parks and other forms of protected areas for ecologically sensitive or particularly biodiversity-rich areas provide many examples of this development. The challenge for policymakers and natural resource users across the region is to establish a balance between socio-economic needs and ecological stability. However, the region's environments are not easily managed due to the prevalence of natural hazard events that compound the pressures from economic activities.

'Acts of God' and a helping hand: natural and human impacts

Understanding environmental change requires an appreciation of two interlinking processes. The first relates to natural processes involving long-term climatic change at one end of the spectrum and shorter-term, high-impact events such as natural disasters at the other end, with the latter being referred to by some as 'Acts of God'. The other set of processes refer to the 'helping hand' of human activities, which impact upon the natural characteristics of the environment via intervention. The degree to which the processes are interlinked creates difficulty in terms of assessing the causal factors of environmental change, and their relative importance in any given context.

The extent to which environments can be assessed according to the typology mentioned previously depends on numerous factors. These include the nature of the impacts on the environment, both positive and negative, the intensity and extent of those impacts, the condition of the environment in terms of its stability, sensitivity and recovery capacity, and the time horizon of the impacts – historical, ongoing or sporadic/episodic. An assessment of these factors will determine the ecological sustainability and, in turn, part of the socio-economic sustainability of the human activity that leads to the impact. However, it is not only human impacts such as cultivation, tourism, mining, fishing or road-building that determine environmental outcomes, but also the dynamic features of the natural environment itself.

When one considers the devastation wrought by natural disasters across Middle America, the notion that environmental change is largely determined by human activities is undermined. Instead, we should view human activities as being highly influential in terms of intensity of natural events. Table 8.1 reveals the regularity of natural hazard events in the region, highlighting the impacts of hurricanes and flood events. It is the flood events that are most influenced by human activities on the environment. Impacts could be mitigated by better environmental management in order to prevent a degraded landscape becoming an irrecoverable

Table 8.1 Major natural disasters and human impacts in Middle America, 1992–97*

Type of natural disaster	Central America[a]	Caribbean[b]	Mexico
Hurricane	1	5	5
Flood	8	9	1
Landslide	2	–	–
Earthquake	–	1	1
Storm	2	2	–
Volcanic eruption	2	1	1
Tsunami	1	–	–
People affected	1,482,900	1,995,100	379,000
Deaths	415	1,220	678

Source: Pan-American Health Organisation (1998: 248–249).
Notes
* In this case, a 'major' disaster is measured in terms of the severity of impact on life and the economy, also the degree of international assistance in remediation and reconstruction
[a] Includes Costa Rica, El Salvador, Honduras, Nicaragua, Panama
[b] Includes Antigua and Barbuda, Bahamas, Cuba, Dominican Republic, Haiti, Jamaica, Montserrat, St Kitts and Nevis, St Lucia, Virgin Islands (US)

one (McGregor 1995). The Haiti flood in November 1994, when 1,122 died and 1.5 million were affected, is a good example of this given the country's considerable erosion problems. It is estimated that Haiti loses 40 million tonnes of soil (the equivalent of over 10,000 hectares of arable land) each year due to heavy deforestation, plantation agriculture and peasant agriculture in a densely populated country. It is not unreasonable, therefore, to make a link between land cover changes and the intensity of flood events (Arthur 1996; PAHO 1998). This same vulnerability of degraded natural environments to intense flood events led to the devastation in Central America following hurricane 'Mitch' in late 1998 (see Box 8.1).

A major concern for policymakers throughout the region is that an argument based on population density as the principal factor in environmental degradation does not hold (see also Bradshaw *et al.*, this volume). In contrast to Haiti with its high levels of degradation and high population density, the case of Belize provides an example of a country with low population density but rising environmental degradation. Rather than poverty and high population densities being factors of environmental degradation in isolation, the prevalence of degradation throughout the region prompts the need to address other factors such as land allocation policies, and price and trade policies that affect the selection and intensity of economic activities (López and Scoseria 1996).

What can be stated with considerable certainty is that human activities do influence the impacts of natural events, in the same way that humans adapt their lifestyles and activities to accommodate them. Nevertheless,

Box 8.1 Hurricane 'Mitch': natural hazards and environmental risks

When hurricane 'Mitch' died out at the end of October 1998, about 10,000 people were dead and another 10,000 were missing. The majority of these people were among the poorest sectors of rural society. The torrential rains associated with the hurricane had given rise to flooding and landslides that had carried away or buried people, their land and their houses. Although hurricane events cannot be avoided, the impacts of these events can be reduced with better forecasting and preparation. Changes in land use activities, settlement construction and disaster mitigation strategies can all limit loss of life and reduce the longer-term impacts on development. Central to the intensity of the hurricane 'Mitch' impacts were the ways in which heavy rains led to landslides and rivers breaking their banks. There is little debate that the capacities of the local environments to manage the storm conditions, i.e. infiltration rates, soil stability and natural land cover, had been negatively affected by agricultural practices over several decades. Denuded hillsides and exposed land surfaces (for large-scale agriculture) accelerated rather than slowed the water flow and its related impacts. For example, the villages located around the Casitas volcano were buried by mudslides from the unstable, deforested hill slopes; an estimated 2,000 people died in this single area. Despite the rapid organisation of civil society, such as the 320 organisations that formed the 'Civil Co-ordinator for Emergency and Reconstruction' in Nicaragua and the international Stockholm Summit (May 1999) on the disaster which agreed a figure of US$9 billion for reconstruction and development, the long-term damage from soil erosion and river sedimentation will affect the recovery process and the ability of many communities to return to their former rural livelihoods.

Indicators	Honduras	Nicaragua	Guatemala	El Salvador	Total
Dead	7,007	2,863	268	240	10,378
Missing	8,052	948	121	19	9,140
Affected	4,753,537	867,752	734,198	346,910	6,702,397
Total Pop.	6,203,537	4,492,700	11,645,900	6,075,536	28,417,324

Sources: Boyer and Pell (1999); Campodónico and Valderrama (2000); Bradshaw *et al.* (this volume).

the frequency of natural disasters and the extent of many human-induced environmental impacts suggest that risks are not being reduced, but rather enhanced by current economic activities. In order to address the interdependence of natural and human impacts, it is imperative to situate environmental change issues in their political context, since it is in the realm of politics (local and national) that strategic decisions about environmental protection, economic development and poverty alleviation must be taken.

Plate 8.1 Deforestation in Chalatenango, El Salvador
Photo: Cathy McIlwaine

Political ecology: the politics of environmental change

The importance of the political context of environmental change, as opposed to more technical, natural science approaches, has been highlighted during the 1990s by geographers using a 'political ecology' approach to environmental change analysis (see Bryant and Bailey 1997; Barton 1999a). This approach focuses on actors, agenda-setting and decision-making issues in environmental change. The ways in which environmental agendas are set, negotiations and decision-making are organised, and how implementation is managed varies considerably between countries. Factors that determine these issues include the institutional structures in place and their capacities and resources, the range of stakeholders or interest groups involved, political will (usually determined by the degree of environmental degradation and the level of public awareness), and the economic context.

During the twentieth century, the economic context was most influential in terms of shaping the region's environments. In particular, the interests of agribusiness and traditional elites led to the slow development of the environmental agenda during the late twentieth century. More often than not, environmental policies have been perceived as a burden on economic activities, and an obstacle to national competitiveness and economic development. This perspective predominated within government and business circles, and environmental issues were often legislated for but poorly supported due to under-resourced and politically weak institutions; this situation led to inadequate monitoring and enforcement of legislation. A lack of institutional support was also an outcome of a strong reliance within these circles on the capacity of business activities to adapt to environmental change via technology and improved environmental management rather than regulation and state intervention; this approach can be referred to as weak ecological modernisation (see Barton 1999b).

As a result of economic prioritisation, the environments of the region have been under-protected, leading to conditions of environmental crisis in some cases, and serious threats elsewhere. Social crisis has often accompanied these processes, particularly in the cases of indigenous groups and the rural and coastal poor. For example, Colchester (1997) cites the case of the 60,000 Amerindians in the interior of Guyana who have been further marginalised by the sale of mining and logging concessions in their traditional territories. The international context of broader development processes presents an additional aspect, such as the impact of Mexico's entry into NAFTA and the apparently positive impact of the side-agreement on environment in terms of its effects on Mexico's domestic environmental policy (Husted and Logsdon 1997). Although the Commission for Environmental Co-operation and the 'Border Plan' for US–Mexico border environmental control, which emerged from the side-agreement, are designed to improve environmental protection and to encourage the enforcement of national environmental laws (see Benton 1996), the positive impacts of a new institution for environmental governance may do little to outweigh the environmental pressures from an intensification of export-manufacturing and agricultural production. It is within this international political economy context that decisions regarding national environments and environmental policies have been established.

A key issue in the debate over power relations and environmental decision-making is the extent to which poverty has forced poor people to exploit marginal lands beyond sustainable levels. It is clear that concentration of land ownership, increasingly capital-intensive agriculture and vulnerability to international market changes have reinforced the predicament of the rural poor, encouraging environmentally degrading activities. National policies have also been influential in providing a disincentive to those farmers producing for local or regional markets (Acevedo *et al.* 1995). Rural livelihoods remain worse than those in urban areas when measured in terms of health indicators (Table 8.2). There is also the anomaly of food security. Although the region has concentrated on food exports, oriented primarily towards North America, particular countries have experienced a reduction in the average daily per capita calorie supply (kilocalories, 1992–94 compared with 1982–84: WRI 1998–99): Cuba, Dominican Republic, Haiti, Nicaragua, Trinidad and Tobago. The reasons behind these reductions vary, but the factors include conflict, international trading restrictions, low agricultural productivity and shifts towards export-orientation in agriculture.

Table 8.2 Urban–rural differences in environmental health in selected Middle American countries (percentage of population with access to safe drinking water[a] and adequate sanitation[b])

Country	Safe drinking water, 1990–96[c]			Adequate sanitation, 1990–96		
	Urban	Rural	Total	Urban	Rural	Total
Costa Rica	100	92	96	95	70	84
Cuba	96	69	89	95	82	92
Dominican Republic	80	–	65	76	83	78
El Salvador	85	46	69	91	65	81
Guatemala	87	49	64	73	52	59
Haiti	37	23	28 ·	42	16	24
Honduras	96	79	87	97	78	87
Jamaica	–	–	86	100	80	89
Mexico	92	57	83	85	32	72
Nicaragua	84	29	53	77	34	60
Trinidad and Tobago	99	91	97	99	98	79

Source: World Resources Institute (1998: 250–251).
Notes
[a] Includes treated surface water, untreated water from protected springs, boreholes and sanitary wells
[b] Refers to adequate excreta disposal facilities that can effectively prevent human, animal and insect contact with excreta
[c] Refers to the most recent year available within the range given

Exhausting work: agriculture, ranching and forestry

Plantation agriculture in Middle America has characterised the regional economy since the Caribbean sugar plantations of the British and French empires and the more recent US fruit (mainly banana) plantations of Central America (see also Thorpe and Bennett, this volume). The environmental impacts have been profound and have intensified during the late twentieth century in the drive for production intensification and in response to stricter buyer and consumer requirements in export markets, for example phytosanitary regulations and product quality.

The greatest threat to natural environments has been the removal of natural vegetation and exposure of soils, leading to a range of degrading processes: water erosion, wind erosion, chemical deterioration (salinisation, acidification, pollution), and physical deterioration (compaction, water-logging, subsidence of organic soils) (WRI 1992). Middle America can be regarded as a region in crisis in terms of these processes, and the only region in the world where agricultural activities rank higher than any other factor in terms of contribution to erosion (see Furley 1996; WRI 1998). The underlying reasons for high rates of soil erosion are to be found

Table 8.3 Human-induced soil degradation, 1945–1990

Region	Total degraded area (million ha)	Degraded land, percentage of vegetated land
Central America and Mexico		
Total degraded area	62.8	24.8
Moderate, severe, extreme	60.9	24.1
Light	1.9	0.7
South America		
Total degraded area	243.4	14.0
Moderate, severe, extreme	138.5	8.0
Light	104.8	6.0
North America		
Total degraded area	95.5	5.3
Moderate, severe, extreme	78.7	4.4
Light	16.8	0.9
World		
Total degraded area	1,964.4	17.0
Moderate, severe, extreme	1,215.4	10.5
Light	749.0	6.5

Source: World Resources Institute (1992: 112).

in the agricultural activities practised by commercial enterprises and, as a consequence of commercial agricultural extensification and dislocation, the activities of smallholding and peasant farmers.

The need to reduce soil erosion in the region can be regarded as the highest priority for environmental agencies and agricultural institutions. Without adequate safeguards, soil erosion rates will continue at current or increased levels and lead to soil productivity losses, and sedimentation and eutrophication of waterways and reservoirs (Alfsen *et al.* 1996). Between 1945 and 1990, Middle America registered the highest percentage area of soil-degraded land in the world; within this percentage, it registered the highest level of moderate, severe or extreme degradation (Table 8.3).

There are two policy directions required to shift away from the unsustainable agricultural practices that operate across the region. One is the targeting of the weak environmental management techniques of commercial organisations in order to raise awareness of impacts and to promote more responsible agricultural practices, such as soil erosion reduction strategies and decreased chemical inputs. The other is the support of agricultural activities of poor rural communities to assist them in managing their activities more sustainably. This requires policies directed at land tenure issues, agricultural input and output pricing, and improved information about farming techniques. Of these factors, the land tenure issue would appear to be at the heart of the matter since most degradation by poor farmers and wood fuel users is carried out on marginal lands that are inappropriate for their activities. In order to protect marginal lands, the ownership and management of productive land must be reconsidered by legislative bodies (see Faber 1993; Utting 1994). Inevitably, the land issue is a highly contentious area of debate. In Nicaragua, for instance, the 1991 Strategic Framework for Agrarian Reform overturned the 1981–88 Sandinista agrarian reform programme, leading to large numbers of claims, primarily from former combatants on both sides (Contras and Sandinistas), as well as pre-Sandinista landowners (Abu-Lughod 2000). However, despite the political complexity of property rights issues, the failure to engage with them in a purposeful way is to fail to engage with the underlying roots of poverty and environmental degradation in the region.

Considerable advances have been made in two areas in the region in terms of improving conventional (high yield emphasis) agriculture and its environmental impacts; both areas fit into a definition of agro-ecology. Agro-ecology involves the integration of agricultural production with environmental protection, primarily by attempting to link the demands of production with as little damage as possible to the natural ecosystem, and to move beyond this in order to generate positive ecological benefits from the preservation of the natural environment, such as pest control and soil fertility outcomes. One area is the development of 'shade' coffee (see Box 8.2). The other is the promotion of organic farming methods, such as in Cuba, where there is extensive recycling of food and animal waste for feeds and fertilisers, and the promotion of a national biotechnology industry to develop biofertilisers and biopesticides (Rosset 1996).

The recognition of the environmental degradation caused by plantation monoculture has led to a return to more 'biodiversity-friendly' methods of production. Instead of removing natural vegetation and planting crops as the only species, there is an attempt to use natural vegetation as a resource for the agricultural system, to act as protection for the crop from intense climatic conditions, from pests which thrive on monoculture (due to the elimination of the habitats of their predators), and from soil instability. Under these conditions, the opportunity to move away from heavy chemical inputs such as pesticides and fungicides are increased.

Although agriculture has been at the centre of the region's development pattern since the colonial period, the environmental problems in rural areas have been intensified by the post-Second World War growth in cattle ranching. Its growth in Central America is closely related to US domestic meat consumption and the rise in fast-food consumption in particular; Bill Weinberg (1991) calls this the 'hamburger connection' and makes the point that this (US agricultural interests) was an important factor in the US low-intensity conflict in the region during the 1980s. The environmental problems associated with ranching, particularly the removal of natural vegetation for pasture and the compaction of the soil by the cattle, make it a low productivity activity, yet this has not deterred governments and business interests from promoting it.

In Costa Rica, national legislation during the 1970s gave land ownership rights to people who cleared land and used it for one year, which often led to the sale of the land to cattle ranchers after that year had expired. By 1984, half of Costa Rica's agricultural land had been converted to pasture despite land use studies that revealed that two-thirds of it were

Box 8.2   Coffee: towards 'biodiversity-friendly' production

Coffee is one of the principal traded commodities in the global economy, with two-thirds of the product emanating from Latin America and the Caribbean. During the 1950–90 period, coffee production in the region increased by 196 per cent, the bulk of it in 'unshaded' plantations. In Mexico, this approach was promoted by the state agency for coffee production and trade (INMECAFE). The environmental impacts of plantation agriculture can be separated into traditional impacts, such as removal of natural vegetation cover for extending plantations (exposing soils to erosion), reductions in biodiversity and soil exhaustion, to modern impacts associated with pesticide use and fertiliser inputs to protect the crop and increase productivity (which have impacts on both the ecosystem and on workers' health).

The main issue is that of the shift from 'shade' coffee grown alongside fruit trees and other hardwoods (a polyculture coffee agro-ecosystem), to 'unshaded' coffee grown following natural vegetation clearance; unshaded coffee currently accounts for approximately 40 per cent of production in Middle America. The biodiversity losses that have accompanied the shift are best identified in terms of the bio-indicator of bird species diversity – it is estimated that unshaded coffee environments support 90 per cent fewer species than shade coffee environments. Although unshaded coffee fulfils the criteria of conventional agriculture – high yields from intensive chemical inputs – it fails in terms of environmental sustainability. The role of international environmental consciousness and the ways in which this has affected consumer choices in developed economy markets will be important in reversing the shift to unshaded coffee. The ability to certify organic coffee and promote small-scale production via international 'fair trade' organisations offers a lifeline to shade coffee producers who receive a larger share of the premium prices and are therefore able to act as environmental stewards in maintaining biodiversity and reducing soil degradation.

Sources: Nestel (1995); WRI (1998); Gobbi (2000).

unsuitable and should have remained forest (WRI 1994). The use of 'slash and burn' or shifting cultivation in the process of land clearance was also an important contributory factor to ecological degradation due to soil erosion and smoke from intentional burning. The gravity of the situation in Costa Rica gave rise to a shift in policy under the administration of President Arias from 1988, leading to greater resource protection via national parks and ecotourism, and the attraction of international finance such as debt for nature swaps to support sustainable development promotion. Approximately 25 per cent of the country's land area has been set aside for conservation or protected areas, building on the process of 'green' development that had seen visits to seven major national parks increase

Table 8.4 Deforestation and land use (percentage land area) in selected Middle American countries

Country	Deforestation[a]		Arable[b]		Cropland[c]		Other[d]	
	1981–1985	1990–1995	1980	1996	1980	1996	1980	1996
Costa Rica	3.6	3.0	5.5	5.6	4.4	4.8	90.1	89.6
Cuba	0.1	1.2	23.9	34.3	6.4	6.8	69.7	58.9
Dominican Republic	0.6	1.6	22.1	27.9	7.2	11.4	70.6	62.4
El Salvador	3.2	3.3	27.0	30.7	8.0	10.5	65.0	59.9
Guatemala	2.0	2.0	11.7	12.6	4.4	5.1	83.9	82.4
Haiti	3.8	3.4	19.8	20.3	12.5	12.7	67.7	67.0
Honduras	2.3	2.3	13.9	15.1	1.8	3.1	84.3	81.9
Jamaica	3.0	7.2	16.6	16.6	5.5	6.1	77.8	79.8
Mexico	1.3	0.9	12.1	13.2	0.8	1.1	87.1	87.0
Nicaragua	2.7	2.5	9.5	20.2	1.5	2.4	89.1	78.8
Panama	0.9	2.1	5.8	6.7	1.6	2.1	92.5	91.1
Puerto Rico		0.9	5.6	2.5	5.6	4.4	88.7	91.3
Trinidad and Tobago	0.4	1.5	13.6	14.6	9.0	9.2	77.4	76.2
Latin America/ Caribbean	1.5/0.1[e]	0.6	5.8	6.8	1.1	1.3	93.1	92.3
World	0.3	0.3	10.1	10.6	0.9	1.0	88.7	88.3

Sources: World Resources Institute (1992: 286); World Bank (1999: 120–122).
Notes
[a] Average annual deforestation (percentage change)
[b] Under temporary crops, fallow or meadows, including pasture, horticulture
[c] Permanent cropland, including shrubs and trees not grown for wood timber
[d] Forest and woodland, uncultivated land, grassland not for pasture, wetlands, built-up areas
[e] 1.5: Central America and Mexico; 0.1 Caribbean subregion

by 50 per cent between 1986–88 (Budowski 1992; UNDP 1998). The considerable drop in deforestation rates in the 1990s compared with the 1980s reveals that a positive impact has been made (see Table 8.4).

The impacts of non-traditional exports

Although there are environmental risks inherent in the plantation and ranching models of agricultural development in the region, the introduction of non-traditional export products (NTAX) such as flowers and high value fruit and vegetables in an attempt to raise foreign exchange earnings during the 1980s has not reduced the risks, but merely shifted the emphasis (see Barham *et al.* 1992; also Thorpe and Bennett, this volume). The rise in seafood production from aquaculture (farmed fish and shellfish) rather than capture fishery (wild fish) sources has also posed serious environmental problems (see Box 8.3). While the risks from traditional agriculture

Box 8.3 Shrimp production and mangrove destruction

Apart from the agricultural products that constitute the NTAX sector, there has also been rapid growth in seafood production. The crisis of capture fisheries around the world has led to the promotion of aquaculture – fish and shellfish farming – in many regions. In Latin America, Ecuador and Central American countries have focused on shrimp as non-traditional products for export to the markets of Asia and North America. Although shrimp is a higher value per unit product than most traditional exports, the environmentally degrading consequences of shrimp farming are also high. The principal concern is the destruction of coastal mangroves in order to establish the shallow, tidal areas for shrimp cultivation; in Nicaragua an estimated 60 per cent of mangroves (500 sq. km) had been destroyed by 1991. As shrimp exports gathered pace during the 1990s, more mangroves were felled to expand these production sites. What is lost in the process is not only the mangroves themselves, but a biodiverse coastal ecosystem that is sustained by the protection offered by the mangrove root systems and the nutrients supplied by the tidal flow.

Sources: Dewalt *et al.* (1996); Stonich (1991); Stonich *et al.* (1997).

were well known, such as deforestation and loss of biodiversity, soil exhaustion, soil erosion, and siltation of surface waters, the risks from non-traditional agriculture are more closely linked with human health since, in the context of NTAX, the high value of the products and the shift from extensive to intensive production strategies have involved the increased use of chemical inputs.

Alongside their desired effects on crop protection and productivity, there are serious undesired effects from chemical inputs, such as pesticide residues in the crop, impacts on natural flora and fauna, and workers' health. Due to its effective isolation from the capitalist world economy throughout the 1980s, agricultural practices and trade patterns in Nicaragua were wholly different from those of its neighbours. Indeed, there is evidence to suggest that general levels of environmental degradation remained relatively low during the period of the Sandinista administration (1979–90), although not in the case of mangroves (see Box 8.3). Cuba has a similar story of lower environmental degradation due to its isolation, with pesticides usage dropping by 70 per cent during the 1980s and biological controls promoted instead (WRI 1994; Gibson 1996).

The threat to the natural environment and workers from pesticides and heavy fertiliser use is of growing concern in the region. As with all conventional agriculture, the late-twentieth-century growth in production levels has been closely linked to increased inputs and controls. For workers, who may suffer from the cumulative effects of use and risks from poor practices

Table 8.5 Reported cases of acute poisoning from pesticides in selected Middle American countries

Country	1993	1994	1995	1996
Nicaragua	—	799	1,207	1,128
Costa Rica	382	583	989	792
Guatemala	282	237	—	—
El Salvador	—	—	1,961	1,469

Source: Pan-American Health Organisation (1998).

(for example, spraying techniques), there is concern that a shift to non-traditional products will increase vulnerability since these crops are most often of higher per unit value and are therefore considered to be worth more intensive protection; for example, agrochemicals may account for up to 35 per cent of production costs (Thrupp 1996).

In Guatemala, a study of pesticide residues in mothers' milk confirmed cases where pesticide residues were 250 times higher than that permitted in cow's milk (Murray 1994). Women in particular are at high risk due to their concentration in the NTAX sector, such as in the export flower business. Previously men working in plantation agriculture were the highest risk group: in Costa Rica in the 1970s, 1,500 men working on banana plantations became sterile as a result of repeated contact with a pesticide (WRI 1994). With pesticides production and imports increasing during the 1990s, for example a 24.8 per cent annual growth rate in Nicaragua (1990–95), and evidence of mismanagement of the products, there are serious concerns about acute poisoning and the longer-term health of workers and their children (Thrupp 1995; Pan-American Health Organisation 1998; see Table 8.5). A further issue relating to pesticide use is the development of resistance – the 'pesticide treadmill' – which has led to pest infestations in certain regions and among particular crops.

Pesticide use also has implications for trade, since detection of pesticide residues in export products can lead to refusal of entry into the intended markets. In the case of the US, which is the primary market for Central American agricultural exports, strict regulations relating to tolerance standards led to 14,000 detentions of non-traditional agricultural exports from Latin America and the Caribbean between the mid-1980s and mid-1990s, to a value of US$95 million (Thrupp 1996; see also Thorpe and Bennett, this volume). To move beyond the high-input, high-risk conventional agriculture methods will require a range of more eco-friendly options such as integrated pest management (IPM) involving plant breeding strategies (to increase resistance), natural pest parasite use, and inundative releases (for example, of wasps, which are egg parasites of many pests) (MacKay 1993). Many of these options may indeed be lower-cost alternatives compared with current methods; for example, in the case of soil fertility, a study of Honduran peasant farmers revealed that the use of velvet bean as a nitrogen-fixing crop led to a reduction of 22 per cent in costs related to chemical inputs (Mausolff and Farber 1995).

Plate 8.2   Environmental destruction due to bauxite mining, Jamaica
Photo: Cathy McIlwaine

Development sanctuaries: the creation of protected areas

The vast majority of agriculture and ranching activities are carried out on land that has been converted from natural vegetation cover, principally forest. The shifting patterns of land use, towards pasture and permanent cropland at the expense of other forested and grassland areas can be observed as a general trend across the region, leading to deforestation pressures (see Table 8.4, page 202). Not only have forests been heavily logged or felled to clear land for pasture or agriculture, there are also issues associated with wood fuel use and the introduction of exotic species. For example, in the Dominican Republic, where 30,000 hectares of forest were lost each year during the 1980s (leaving only 1 million hectares by 1999), a fast-growing tree (*acacin mangium*) was encouraged as a timber cash crop. Despite its commercial advantages, it has posed a serious threat to species diversity, and has challenged traditional land use patterns and competed with kitchen garden cultivation that has traditionally been important for food security (Sagawe 1991; Rocheleau and Ross 1995).

A response to the environmental degradation of large tracts of natural vegetation by commercial and small-scale economic activities has been the creation of protected areas ranging from small, regulated areas of natural vegetation to internationally recognised areas. Middle America now has 21 Biosphere Reserves, 5 (natural) World Heritage Sites, and 33 Wetlands of International Importance (WRI 1998; UNESCO 1998). The nature and extent of these areas vary considerably across the region (see Figure 8.1). At the other end of the spectrum one can find the smaller conservation

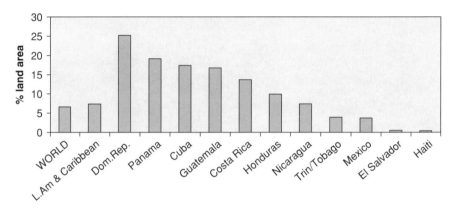

Figure 8.1 National protected areas in Middle America, 1996
Source: World Bank (1999: 132–134).
Note
National protected areas according to the World Conservation
Monitoring Centre include (a) scientific reserves, or strict nature
reserves with limited public access, (b) national parks of national or
international significance, (c) national monuments and natural
landscapes with unique aspects, (d) managed national reserves and
wildlife sanctuaries, (e) protected landscapes and seascapes. Areas
a–c are totally protected, d–e are partially protected.

projects that target habitat protection in order to protect threatened species. In the case of Guatemala, the National University of San Carlos in
Guatemala City has set up *biotopos* (small reserves) to achieve this, such as
the programme to protect manatees in Lake Izabal, and another to protect
the quetzal bird (the national symbol) in Baja Verapaz (Weinberg 1991).

In many ways, the creation of national parks and designated ecological
areas within which activities are regulated has been a distraction from the
environmental impacts of the economic development model. Rather than
challenging the model and addressing the degrading activities themselves,
the separation of environments into relatively unprotected economic zones,
and highly protected ecological zones, has been a feature of recent national
environmental protection strategies. In terms of conservation objectives of
protecting biodiversity and natural environments from damaging human
interventions, the setting-up of protected areas has been constructed as a
way of slowing the unchecked expansion of deforestation for wood products, agricultural land and ranching, and of establishing opportunities
for new economic activities such as ecotourism. However, it is clear that
this is not a panacea for environmental degradation in the region. Only if
protected areas are established within the context of changing activities
beyond the protected areas, and in collaboration with all affected stakeholders, can they be effective. One principal objective of protected areas
is to encourage local communities to reduce their dependence on forest
products, especially those that are derived in an unsustainable way.

The drive towards establishing protected areas became especially strong during the late 1980s, following the recognition of the implications of rapid forest depletion not only in terms of the forests themselves but in terms of the consequences for soil stability, river sedimentation and vulnerability to natural hazards. By the late 1990s there had been a surge in national park creation, for example in Jamaica, the Montego Bay Marine Park, the Blue and John Crow Mountains National Park, the Cockpit Country and Black River wetland. Nevertheless, the problems associated with regulating these spaces and controlling activities remain. In the case of the Jamaican parks, David Smith notes that the slow development of regulations by the Natural Resources Conservation Authority Act, the reluctance of judges to impose realistic fines, and a lack of park patrolling resources all work to the detriment of the effective management of the parks (Smith 1995).

In Central America, following the conflicts of the 'lost decade' of the 1980s, the proposal for bilateral or multilateral regional approaches to protected areas that would straddle national boundaries was put forward as a way of achieving better national and environmental integration. These were given the title 'Peace Parks' and, beyond their political aims of improving regional relations, were realistic approaches to managing ecosystems according to their natural delimitations rather than within national boundaries (Arias and Nations 1992). In the case of Costa Rica, the management issue has been taken seriously, with a co-ordinating environment ministry, a National Biodiversity Institute (INBIO), decentralised natural conservation units, and 36 natural resource vigilance committees that involve over 3,000 citizens (Umaña and Brandon 1992; UNDP 1998). However, the Costa Rican example is not matched across the region, suggesting that environmental management issues are not taken as seriously as they might be, especially in terms of resourcing.

Cleaning up coastal development

The Caribbean coasts of Middle America were developed rapidly during the late twentieth century as tourist destinations, creating new pressures for the coastal environments and a need for more integrated and effective management. David Fennell (1999) describes these pressures as 'stressor activities': permanent environmental restructuring (such as construction activities and land use changes); generation of waste residues (increasing local pollution); the tourist activities themselves; and lastly, the effects on local population dynamics due to the strong seasonality of the sector.

The example of Cancún in Mexico provides the most extreme case study of this phenomenon. Following the increase in tourism activity along the 'Mexican Riviera' (Acapulco and Puerto Vallarta) during the 1950s and 1960s, the Mexican government developed a plan in 1969 to build a vast tourism complex on the southeast coast. Construction was rapid and little attention was paid to the mangroves, coastal dunes and internal lagoon in the area. It was typical of Latin America's prestige projects of the period, a mega-project designed to overcome nature rather than integrate with it.

By the mid-1990s, the resort had 20,000 hotel rooms within a city of tourists and approximately 80,000 tourism-related workers (Simon 1997). An area of low population density and sustainable coastal livelihoods had become the largest tourist destination in Latin America in less than 20 years, generating foreign exchange and local jobs but at the expense of environmental quality.

Mass tourism on Mexico's east coast and on the larger Caribbean islands has definitely increased pressure on coastal environments. The impacts can be categorised as residential and structural, and overlap with the stressor activities. On the residential side, the population increases have outpaced the infrastructure capacities for water, wastewater and solid waste management. This has led to difficulties in terms of providing potable water, but more importantly in terms of untreated sewage being channelled into the sea and the expansion of poorly managed or illegal landfills. The environmental repercussions include a deterioration of seawater quality, which affects marine life feeding and reproduction and the health of bathers, and public environmental health concerns in terms of rodents, pests and other vectors of disease transmission. The solid waste management issue came to a head in Barbados in the mid-1990s when the government proposed the construction of a landfill facility in a designated national park (Beckles 1996). The impacts of all these environmental problems are not felt equally by all local interest groups. A study by Susan Stonich (1998) of tourism development on the Bay Islands of Honduras concludes that it is the poor *ladino* immigrants and local Afro-Antillean residents who are most vulnerable to environmental health risks from tourism development.

Underlying these issues is the essential structural debate around the environmental impacts associated with construction and infrastructural developments (roads, airports, etc.) for tourism development. A further structural issue is the impact of the leisure pursuits of the tourists in these destinations (see Box 8.4). Although there should be strategic thinking about the sustainability and tourism-carrying capacities of different locations, the implementation of best practices by hotels and other tourism agencies can reduce impact levels considerably, for example, erosion and sediment control, selective purchasing to reduce packaging, low-flush toilets and low-flow shower heads, reuse of grey water for irrigation, and energy conservation (Vanzella-Khouri 1998). Education, in the form of business and community awareness programmes, is also a vital area for development by NGOs and government.

The emergence of high value-added tourism that mirrors many of the advantages of ecotourism in forested areas can be seen in Caribbean reef-diving activities. A further example is that of whale-watching off the coast of Mexico's Baja California (Young 1999), which acts as an ecotourism opportunity for local communities. Nevertheless, without protection of the tourist attraction, revenues will dimish rapidly. Reefs, for example, suffer both direct pressures from overfishing, destructive fishing practices (such as fishing-gear damage), as well as effects related to climate change (for instance, coral bleaching, where coral algae abandon the reef, leading to the death of the coral, has been revealed as a response to sea temperature rises associated with global warming), and contaminated urban and agricultural runoff and sedimentation (see Wilkinson 1996). The conservation of the resource brings a wide range of stakeholders into contact, and often conflict: conservationists, divers, tourism operators and local fishers who depend on artisanal, inshore fisheries techniques, such as line-and-hook, nets and pots, and spear-fishing (see Box 8.4).

Box 8.4 Coral reef degradation and marine protected areas

The establishment of the International Coral Reef Initiative in 1995 marked an important starting point for the protection of coral reefs around the world. One widely favoured measure to combat coral reef degradation is the creation of marine protected areas (MPAs), which combine planning, management, legal protection and law enforcement, and environmental education with a view to reducing degrading impacts. An MPA establishes an area within which certain activities are regulated or excluded. The area is managed by wardens who enforce the regulations, and supervised by an MPA authority that strives to involve a range of stakeholders in the implementation of an acceptable management strategy. In the wider Caribbean, there are 250 MPAs already established or proposed in order to diminish the percentage of reefs (estimated at 60 per cent) considered 'at risk'. The objectives of MPAs include the protection of the reef habitat from destructive extractive activities (for example, dynamite fishing), developing ecotourism, providing more effective fisheries management, providing sites for research, and involving local communities in management and stewardship.

Since there are wide variations in existing degradation, potential risks and user-group categories in different locations, MPAs should be flexible and in certain cases might require a 'no-take zone' where fishing is prohibited in order to protect fish numbers and diversity and also to prevent damage from fishing gear (for example the dropping of anchors). The problems that then arise are those between the MPA authorities and the local fishers who generate all or part of their income from fishing in these areas, and who subsequently must reduce their fishing effort, change their fishing techniques, seek fishing opportunities further afield or stop fishing altogether. For this reason, the marine ecology issues need to be closely tied to wider sustainable development issues in order for them to be successful in both coral reef and human terms. In view of the fact that coral reefs are not only important for fish but also as a coastal buffer against wave and storm impacts, the urgency of protection has been increasingly emphasised.

Sources: Agardy (1998); Vanzella-Khouri (1998); WRI (1998).

The carrying capacity of paradise: from mass tourism to ecotourism

Tourism has increased throughout the region during the 1990s following the political instability of the 1980s, and provides an opportunity for economic diversity and the promotion of lower-impact activities, depending on the scale, suitability and management of the tourism activities. Particular countries, such as the Dominican Republic, Costa Rica and Cuba, have made strategic decisions to promote tourism as a development path, while others have continued to expand the sector gradually as a percentage of export earnings (see Figure 8.2; also Willis, this volume). A fundamental question left unresolved, however, is the carrying capacities of different tourism activities in particular locations. The principal issue is the extent to which tourism development can increase without sufficient controls and safeguards, or viewed from another perspective, the ability of alternative tourism such as ecotourism, to support environmental policymaking and create a 'win–win' scenario of strong export earnings and strong environmental protection.

The environmental problems associated with mass tourism have generated reactions from tourists, agencies and local communities. They reveal the need for a range of different stakeholders to be involved in social, economic and ecological decision-making in order to convert tourist development into sustainable development (Pattullo 1996; Olsen 1997); these decisions include those of ownership of tourist facilities and control of tourist numbers (France and Wheeller 1995). In many cases the reaction has been to seek an alternative path to income-generation, based on scenic beauty and biodiversity in preference to Caribbean 'sand and sea'. However, the shift towards higher-value tourism may in itself create different

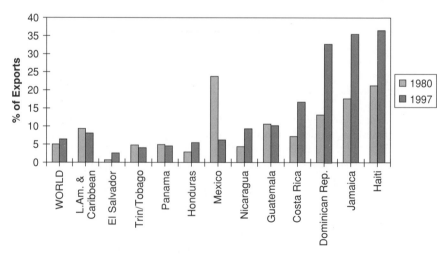

Figure 8.2   International tourism receipts as a percentage of exports
Source: World Bank (1999: 366–368).

environmental problems, such as the pressure for golf courses and more secluded developments in areas of originally high environmental quality (ironically the key locational factor). For these reasons, each tourism project, whether small-scale, high-value or more conventional mass tourism, must be considered *prior* to development on its own merits, within its environmental context in terms of a wide range of sustainability criteria.

Lower-density, lower-impact tourism is already gaining ground as an alternative to mass tourism and can offer more sustainable development opportunities in Middle America, especially for smaller communities and islands where space for development is restricted. The example of Montserrat with the development of a trails network and low-density residential tourism is a model for other small island economies within the Caribbean (Weaver 1995). By promoting tourism based on biodiversity and tranquillity, there are strong incentives to provide a contrast to mass tourism. Belize and Dominica are examples of similar smaller states that have moved directly into ecotourism as a strategic development opportunity. A major opportunity is that the alternative tourism market is quite diverse and different islands and regions can focus on different tourist groups, such as scientific tourists (scientists and student researchers) or one of the three groups of nature tourists: 'hard-nature' (naturalists); 'soft-nature' (nature and wildlife observers with little knowledge); and 'adventure' (outdoors activity-oriented) (Budowski 1992).

The move towards sustainable tourism should seek to achieve three objectives (Lindberg *et al.* 1996): generation of financial support for protected area management; generation of local economic benefits; and generation of local support for conservation. In their own research in Belize to follow up these objectives, Lindberg *et al.* considered that the second and third are generally achieved but that the first objective requires the implementation of user fees. Local community considerations draw out complex issues around development and sustainability of livelihoods, but without full appreciation of social imperatives it is unlikely that many ecotourism projects will be converted into genuinely sustainable tourism activities (see Belsky 1997; Mowforth and Munt 1998; Wall 1997).

Conclusion: towards sustainability

The region's predominant economic model of plantation agriculture, ranching and forestry exploitation has been complemented during the late twentieth century by non-traditional agricultural production and tourism. Rather than reducing environmental risks, the increased use of chemical inputs has compounded pre-existing problems of traditional activities such as deforestation and soil erosion. In summary, the traditional model presented problems of extensification (low productivity per unit area therefore increased demand for deforested land), and the non-traditional model presents problems of intensification (increased inputs for productivity and crop protection per unit area, and concentrated mass tourism). Both models require better environmental management in order to move the region towards greater sustainability.

Awareness of the need for more sustainable economic activities, across a broad range of state, business and civil society organisations, has led to improvements in business practices and regulatory systems. However, levels of enforcement and compliance remain low due to the prioritisation of economic production for export; environmental protection is widely regarded as a trade-off against economic growth and development, rather than an integral part of the process. Although alternative tourism, organic and 'biodiversity friendly' farming, and a shift from large-scale to small-scale activities oriented around community development rather than export earnings present opportunities to diversify away from environmentally degrading economic activities, this diversification can only be successful if accompanied from the outset by strong environmental management systems, such as flexible regulatory regimes and voluntary initiatives. Without safeguards for environmental protection, it is unlikely that any alternative economic strategies will become sustainable options. Sustainability also requires that social needs are also satisfied, and that a wide range of social actors participate in and decide upon economic development opportunities.

The challenge for Middle American governments, in association with a range of stakeholders, including civil society (for example, environmental NGOs and the rural poor), business groups, trading partners, and international organisations (governmental and non-governmental), is threefold: first, to reassess and refocus the insertion of the region into the global economy, in order to move beyond the negative environmental and social consequences of the traditional and non-traditional export models; second, to develop institutional capacities and networks of interest groups in order to manage more effectively the economic demands, social demands and environmental protection needs with a view to sustainable development; and third, to improve environmental controls (via improved participation, monitoring and enforcement) based on existing legislation and international commitments relating to economic activities and health concerns.

The concept of sustainable development is valid and applicable in the Middle American context. With high dependence on natural resource exploitation, mostly export-oriented, the historical development process has been shaped by foreign interests aligned with domestic elite groups operating unsustainable environmental management activities. High rates of soil erosion, loss of biodiversity, increased vulnerability to the impacts of natural hazards, and the 'toxification' of soils, flora and fauna, and workers bear testimony to this process. The task of restoring environments to provide new opportunities for the rural poor and coastal communities, to promote biodiversity, and to reduce environmental risks and disaster impacts is not a simple one, and cannot be achieved via spasmodic financial injections or piecemeal policies. What is required is a long-term process of integrated development. This should focus on education, social development, economic transformation and public policy formulation and implementation, where environment and poverty issues are central rather than peripheral to development discourses and practices.

Further reading

Barker, D. and **McGregor, D.** (eds) (1995) *Environment and Development in the Caribbean: Geographical Perspectives*. University of the West Indies Press, Mona, Kingston.
An edited volume which brings together conference papers from physical and human geographers on coastal management, tourism, natural hazards, land resources and protected areas.

Collinson, H. (ed.) (1996) *Green Guerrillas: Environmental Conflicts and Initiatives in Latin America and the Caribbean*. Latin America Bureau, London.
A collection of short articles that discuss social mobilisation and conflicts relating to environmental degradation, as well as interesting environmental initiatives.

Faber, D. (1993) *Environment under Fire: Imperialism and the Ecological Crisis in Central America*. Monthly Review Press, New York.
A critical assessment of the contribution of US foreign policy and economic interests to Central America's social and ecological 'crisis'.

Simon, J. (1997) *Endangered Mexico: an Environment on the Edge*. Sierra Club, London.
An engaging and accessible account of a wide range of Mexican environmental issues and their political, economic and social contexts.

Utting, P. (1994) Social and political dimensions of environmental protection in Central America, *Development and Change* **25**(1): 231–259.
A useful discussion of the problems of environmental and social policy coherence in Central America, focusing on forestry issues.

References

Abu-Lughod, D. (2000) Failed buyouts: land rights for Contra veterans in postwar Nicaragua, *Latin American Perspectives* **27**(3): 32–62.

Acevedo, C., Barry, D. and **Rosa, H.** (1995) El Salvador's agricultural sector: macro-economic policy, agrarian change and the environment, *World Development* **23**(12): 2153–2172.

Agardy, T. (1998) The role of marine protected areas in coral reef conservation, in Hatziolos, M. *et al.* (eds) *Coral Reefs: Challenges and Opportunities for Sustainable Management*. World Bank, Washington, D.C.: 103–105.

Alfsen, K., de Franco, M., Glomsrød, S. and **Johnsen, T.** (1996) The cost of soil erosion in Nicaragua, *Ecological Economics* **16**(2): 129–145.

Altieri, M. (1995) *Agroecology: the Science of Sustainable Agriculture*. Intermediate Technology Publications, London.

Annis, S. (ed.) (1992) *Poverty, Natural Resources, and Public Policy in Central America*. Transaction Publishers, New Brunswick, NJ.

Arias, O. and **Nations, J.D.** (1992) A call for Central American peace parks, in Annis, S. (ed.) *Poverty, Natural Resources, and Public Policy in Central America*. Transaction Publishers, New Brunswick, NJ: 43–58.

Arthur, C. (1996) Confronting Haiti's environmental crisis: a tale of two visions, in Collinson, H. (ed.) *Green Guerrillas: Environmental Conflicts and Initiatives in Latin America and the Caribbean*. Latin America Bureau, London: 149–157.

Barham, B., Clark, M., Katz, E. and **Shurman, R.** (1992) Nontraditional agricultural exports in Latin America, *Latin American Research Review* **27**(2): 43–83.

Barton, J.R. (1999a) The environmental agenda: accountability for sustainability, in Phillips, N. and Buxton, J. (eds) *Developments in Latin American Political Economy*. Manchester University Press, Manchester: 186–204.

—— (1999b) Latin American environmental management: is ecological modernisation good enough? Paper presented at the Society of Latin American Studies Conference, Cambridge, 9–11 April.

Beckles, H. (1996) Where will all the garbage go? Tourism, politics, and the environment in Barbados, in Collinson, H. (ed.) *Green Guerrillas: Environmental Conflicts and Initiatives in Latin America and the Caribbean*. Latin America Bureau, London: 187–194.

Belsky, J.M. (1997) Misrepresenting communities: the politics of community-based rural ecotourism in Gales Point Manatee, Belize, *Rural Sociology* **64**(4): 641–666.

Benton, J.R. (1996) The greening of free trade? The debate about the North American free trade agreement (NAFTA) and the environment, *Environment and Planning A* **28**(12): 2155–2178.

Boyer, J. and **Pell, A.** (1999) Mitch in Honduras: a Disaster Waiting to Happen, *NACLA Report on the Americas* **33**(2): 36–41.

Bryant, R. and **Bailey, S.** (1997) *Third World Political Ecology*. Routledge, London.

Budowski, T. (1992) Ecotourism Costa Rican style, in Barzetti, V. and Rovinski, Y. (eds) *Towards a Green Central America: Integrating Conservation and Development*. Kumarian Press, West Hartford, CT: 48–62.

Campodónico, H. and **Valderrama, M.** (2000) Latin America: the economy and the environment collapse on the region's poorest people, in Randel, J., German, T. and Ewing, D. (eds) *The Reality of Aid 2000: an Independent Review of Poverty Reduction and Development Assistance*. Earthscan, London: 177–181.

Colchester, M. (1997) Guyana: fragile frontier, *Race and Class* **38**(4): 33–56.

Dewalt, B., Vergne, P. and **Hardin, M.** (1996) Shrimp aquaculture development and the environment: people, mangroves and fisheries in the Gulf of Fonseca, Honduras, *World Development* **24**(7): 1193–1208.

Faber, D. (1993) *Environment under Fire: Imperialism and the Ecological Crisis in Central America*. Monthly Review Press, New York.

Fennell, D. (1999) *Ecotourism: An Introduction*. Routledge, London.

France, L. and **Wheeller, B.** (1995) Sustainable tourism in the Caribbean, in Barker, D. and McGregor, D. (eds) *Environment and Development in the Caribbean: Geographical Perspectives*. University of the West Indies Press, Mona, Kingston: 59–69.

Furley, P. (1996) Environmental issues and the impact of development, in Preston, D. (ed.) *Latin American Development: Geographical Perspectives*. Longman, Harlow: 70–115.

Gibson, B. (1996) The environmental consequences of stagnation in Nicaragua, *World Development* **24**(2): 325–339.

Gobbi, J. (2000) Is biodiversity-friendly coffee financially viable? An analysis of five different coffee production systems in western El Salvador, *Ecological Economics* **33**: 267–281.

Husted, B.W. and **Logsdon, J.M.** (1997) The impact of NAFTA on Mexico's environmental policy, *Growth and Change* **28**(1): 24–48.

Lindberg, K., Enriquez, J. and **Sproule, K.** (1996) Ecotourism questioned: case studies from Belize, *Annals of Tourism Research* **23**(3): 543–562.

López, R. and **Scoseria, C.** (1996) Environmental sustainability and poverty in Belize: a policy paper, *Environment and Development Economics* **1**(3): 289–310.

MacKay, K.T. (1993) Alternative methods for pest management in developing countries, in Forget, G., Goodman, T. and de Villiers, A. (eds) *Impact of Pesticide Use on Health in Developing Countries*. International Development Research Centre, Ottawa: 303–314.

McGregor, D. (1995) Soil erosion, environmental change and development in the Caribbean: a deepening crisis?, in Barker, D. and McGregor, D. (eds) *Environment and Development in the Caribbean: Geographical Perspectives*. University of the West Indies Press, Mona, Kingston: 189–208.

Mausolff, C. and **Farber, S.** (1995) An economic analysis of ecological agricultural technologies among peasant farmers in Honduras, *Ecological Economics* **12**(3): 237–248.

Mowforth, M. and **Munt, I.** (1998) *Tourism and Sustainability: New Tourism in the Third World*. Routledge, London.

Murray, D. (1994) *Cultivating Crisis: the Human Cost of Pesticides in Latin America*. University of Texas Press, Austin.

Nestel, D. (1995) Coffee in Mexico: international market, agricultural landscape and ecology, *Ecological Economics* **15**(2): 165–178.

Olsen, B. (1997) Environmentally sustainable development and tourism: lessons from Negril, Jamaica, *Human Organization* **56**(3): 285–293.

Pan-American Health Organisation (PAHO) (1998) *Health in the Americas, 1998.* PAHO, Washington, D.C.

Pattullo, P. (1996) Green crime, green redemption: the environment and ecotourism in the Caribbean, in Collinson, H. (ed.) *Green Guerrillas: Environmental Conflicts and Initiatives in Latin America and the Caribbean.* Latin America Bureau, London: 178–186.

Rocheleau, D. and **Ross, L.** (1995) Trees as tools, trees as text: struggles over resources in Zambrana-Chacuey, Dominican Republic, *Antipode* **27**(4): 407–428.

Rosset, P. (1996) The greening of Cuba, in Collinson, H. (ed.) *Green Guerrillas: Environmental Conflicts and Initiatives in Latin America and the Caribbean.* Latin America Bureau, London: 158–167.

Sagawe, T. (1991) Deforestation and the behaviour of households in the Dominican Republic, *Geography,* **76**: 333, pt.4, 304–314.

Simon, J. (1997) *Mexico: an Environment on the Edge.* Sierra Club, London.

Smith, D.C. (1995) Implementing a national park system for Jamaica: the PARC project, in Barker, D. and McGregor, D. (eds) *Environment and Development in the Caribbean: Geographical Perspectives.* University of the West Indies Press, Mona, Kingston: 249–258.

Stonich, S. (1991) The promotion of non-traditional agricultural exports in Honduras – issues of equity, environment and natural resource management, *Development and Change* **22**(4): 725–755.

—— (1998) Political ecology of tourism, *Annals of Tourism Research* **25**(1): 25–54.

Stonich, S., Bort, J. and **Ovares, L.** (1997) Globalization of shrimp mariculture: the impact of social justice and environmental quality in Central America, *Society and Natural Resources* **10**(2): 161–179.

Thrupp, L.A. (1995) *Bittersweet Harvest for Global Supermarkets: Challenges in Latin America's Agricultural Export Boom.* World Resources Institute, Washington, D.C.

—— (1996) New harvests, old problems: the challenges facing Latin America's agro-export boom, in Collinson, H. (ed.) *Green Guerrillas: Environmental Conflicts and Initiatives in Latin America and the Caribbean.* Latin America Bureau, London: 122–131.

Umaña, A. and **Brandon, K.** (1992) Inventing institutions for conservation: lessons from Costa Rica, in Annis, S. (ed.) *Poverty, Natural Resources, and Public Policy in Central America.* Transaction Publishers, New Brunswick, NJ: 85–108.

UNESCO (1998) *World Culture Report.* UNESCO, Paris.

United Nations Development Programme (UNDP) (1998) *Human Development Report.* Oxford University Press, New York.

Utting, P. (1994) Social and political dimensions of environmental protection in Central America, *Development and Change* **25**(1): 231–259.

Vanzella-Khouri, A. (1998) Coral reef conservation in the wider Caribbean through integrated coastal area management, marine protected areas, and partnerships with the tourism sector, in Hatziolos, M., Hooten, A. and Fodor, M. (eds) *Coral Reefs: Challenges and Opportunities for Sustainable Management.* World Bank, Washington, D.C.: 209–211.

Wall, G. (1997) Is ecotourism sustainable? *Environmental Management* **21**(4): 483–491.

Weaver, D.B. (1995) Alternative tourism in Montserrat, *Tourism Management* **16**(8): 593–604.

Weinberg, W. (1991) *War on the Land: Ecology and Politics in Central America.* Zed, London.

Wilkinson, C. (1996) Global change and coral reefs: impacts on reefs, economies and human cultures, *Global Change Biology* 2: 547–558.

World Bank (1999) *World Development Indicators.* World Bank, Washington, D.C.

World Resources Institute (WRI)/UNEP/UNDP (1992) *World Resources, 1992–93.* Oxford University Press, New York.

—— (1994) *World Resources, 1994–95.* Oxford University Press, New York.

World Resources Institute (WRI)/UNEP/World Bank (1998) *World Resources, 1998–99.* Oxford University Press, New York.

Young, E.H. (1999) Balancing conservation with development in small-scale fisheries: is ecotourism an empty promise? *Human Ecology* **27**(4): 581–620.

CHAPTER 9

Dependency, diversity and change: towards sustainable urbanisation

Mark Pelling

Introduction

What do the Aztec ruler Montezuma I, the slave leader Toussaint L'Ouverture and US Air Force warplanes have in common? The answer (perhaps not surprisingly) is that they have all played key roles in shaping urbanisation in Middle America. Urban development in the region is characterised by contradiction and extremes, with dramatic and rapid shifts in urban character and the role played by urbanisation in national economic and political life. The region contains both the largest urban area (Mexico City covers 9,560 km^2) and the smallest capital city (Road Town in the British Virgin Islands has a population of 6,000) in the Western Hemisphere. Using the United Nations Human Development Index (UNDP 1999), the region also contains both the most highly and most lowly ranked countries in Central and South America and the Caribbean (Barbados ranked 24 and Haiti ranked 159) (see Willis and McIlwaine, this volume). Amidst such variety we need to ask what form should 'sustainable urbanisation' take and how are its principles being applied in Middle America?

The nature of urbanisation

Until the 1950s few countries in this region had developed urban-based national socio-economies, however, by the end of the twentieth century only Haiti remained predominantly rural in character; 68 per cent of its national population were rural in 1997 (UNDP 1999). As economies move through a development transition they tend to shift from an agricultural towards an industrial and service sector base. Both of these latter activities benefit from the economies of scale and infrastructure offered by urban settlements and consequently national economic growth appears to be closely related to national levels of urbanisation. This pattern is borne out in Middle America, where Haiti, the least urban society, is also the least economically developed nation – GNP per capita was US$380 in 1997 (UNDP 1999). Similar relationships can be found in Guyana (36 per cent urban, GNP per capita US$800) and Honduras (45 per cent urban, GNP per capita US$740) (UNDP 1999). Indicators also suggest that the region's more urbanised societies are more economically developed, although remaining poor compared to economies of the global North. Examples include Trinidad and Tobago (72 per cent urban, GNP per capita US$4,250), Mexico (74 per cent urban, GNP per capita US$3,700) and Dominica (urban population 70 per cent, GNP per capita US$3,040) (UNDP 1999). The way in which planners perceive the relationship between economic growth and urbanisation is critical for shaping urban futures. This is examined in the following section through a discussion of two contrasting perspectives on urban sustainability, called the business-as-usual approach and the ecological city approach respectively.

Since the 1960s, Middle America has had higher levels of urbanisation than either the 'developing world' or the world as a whole. In the insular Caribbean alone, the number of cities with more than 100,000 residents has grown from seven in 1950 to 12 in 1970, and 24 at the end of the 1980s (Potter 1995). Middle American urban systems are characterised by high primacy (the ratio of the population of the largest city to the second city) with up to 60 per cent of national populations living in the capital city. Guatemala has the second highest primacy ratio in the Americas (second to Uruguay), with Guatemala City having 25 times the population of the second city, Quetzaltenango.

The distribution of urban centres is shown in Figure 9.1. Mexico has the most extensive urban system, as you would expect from the region's most populous and largest country. Mexico is the only country to have cities outside the capital with populations in excess of one million. A number of large urban centres are also found in the Central American states and the Greater Antilles, with the Guianas and the Lesser Antilles having small urban centres. However, smallness of urban centre should not be confused with the absence of an urbanised society. Figure 9.2 shows the proportion of national populations resident in urban settlements and the urban growth rate. Some 64 per cent of the region's population is based in urban centres, with the smaller Caribbean islands being the most urbanised (the Cayman Islands are 100 per cent urban, Guadeloupe is 99 per cent, the Bahamas are 87 per cent). On the mainland, only Mexico with 75 per cent has a higher than regional average proportion of its population residing in urban areas.

Urban hierarchies differ widely within individual countries. Some of the smaller Caribbean nations have only a single urban centre (for example, Castries in St Lucia), in others two cities have emerged, often a primate capital city and a supporting port city (San José and Limón in Costa Rica, for example). More developed urban systems are found in countries with more extensive transportation networks, primate cities being supported by a layer of secondary regional centres or ports, as in Jamaica and Guatemala. In Mexico, the region's largest and most integrated national urban

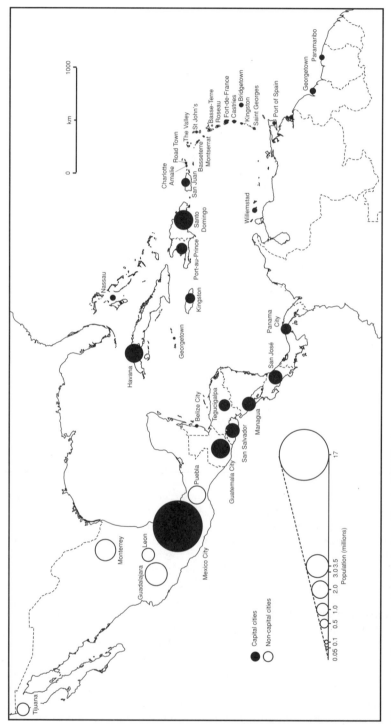

Figure 9.1 Capital and millionaire cities in Middle America
Sources: UNFPA (1999); World Bank (2000).

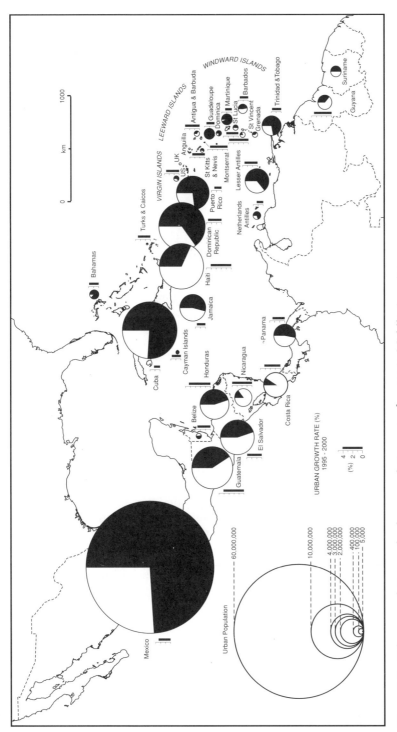

Figure 9.2 Urban population and urban population growth rates in Middle America
Sources: Potter (1995); UNFPA (1999); World Bank (2000).

system (dating to the late sixteenth century), thousands of local centres feed into regional capitals. These in turn support the largest urban areas, dominated by Mexico City and Guadalajara in the south and centre, and border cities like Tijuana and Mexicali in the north. Although it is outside the scope of this book to go into any detail, it should be noted that urban centres outside of Middle America – especially Caracas and Miami – are influential components of the region's urban web, playing important roles as destinations for economic migrants and as sources of cultural and financial returns to Middle America.

The aim of this chapter is to provide an overview of the evolution of cities and city life in Middle America and to take a detailed look at the contemporary challenges for cities attempting to move towards more sustainable urban development. Before this, an outline of what sustainable urbanisation might be, and how it may be achieved, needs to be examined.

Sustainable development and cities

What aspects of cities should be sustained? What should not be sustained? How does sustainable urban development differ from preceding strategies for urbanisation? The most commonly used definition of sustainable development, is that it is 'development that meets the needs of present generations without compromising the ability of future generations to meet their own need' (WCED 1987). But this gives only a little guidance. It says nothing about what kinds of needs should be met or how competing needs within the same generation or between generations can be valued and prioritised. We are no closer to understanding what should be sustained – ecological, social or economic assets, and whether sustainability should be seen as part of a local, citywide, national or global agenda.

A continuum of interpretations of sustainable development has evolved from academic writing and policy statements with two poles of weak and strong sustainability. Weak sustainability is anthropocentric and sees considerable scope for technological solutions to environmental problems. In particular, there is support for the substitutability of natural assets with human capital. Substitutability argues that value is transferable between natural and human assets. For example, turning a forested hillside into a housing development is supported on the grounds that development enhances the monetary value of the land beyond the economic value of the forest. The intrinsic value of the ecological assets and their contributions to environmental regulation (in this case as a carbon sink and slope stabiliser) are seldom considered. Policies associated with weak sustainability favour reform of the present systems of urbanisation. Proponents of strong sustainability hold a more ecocentric view and seek to minimise the loss of natural capital, for example, by preserving urban green space. Substitutability is not supported. Policies promoted by strong sustainability include greater equity in human consumption patterns, with high consumption populations (the wealthy) having to reduce, and low-consumption populations (the poor) being entitled to increase, their consumption levels. There is also a preference for prevention rather than cure; for example, designing cities to reduce transport needs rather than manufacturing fuel-efficient cars.

Much of the literature on cities and development is limited by taking a weak sustainability approach, as in the *World Development Report 1999/2000* (World Bank 2000). This sees cities primarily as engines of economic growth. The role of capital cities in particular is to sustain and encourage the national economy and economic development of the country. In many ways this can be seen as a business-as-usual approach where sustainable development was used to justify trends already well-established in urban development. Support for the business-as-usual approach continues to be widespread, despite mounting environmental and social crises generated by the types of concentrated and extensive urban-economic growth that results. The business-as-usual approach has been challenged by a more ecologically centred perspective (Girardet 1999; Hardoy *et al.* 1992). The urban ecology school criticises the linear metabolism of cities where flows of raw materials (food, water, chemical and human energy, inventiveness) are turned into products (political power, manufactured goods and services, social and biological reproduction) and waste. Goods are drawn in from increasingly distant environments to satisfy the urban metabolism. The distance from which goods are acquired and to which wastes can be sent is linked to the developmental stage of the city. Smaller and poorer cities will tend to draw on local hinterlands, the largest cities will draw upon a global market and contribute significantly to global environmental problems. The worst environmental problems are found in marginalised areas of poor or middle-ranking cities. Here local waste (rubbish, sewerage, etc) accumulates, together with waste from wider urban processes (industrialisation, traffic) and even global trends (global climate change and sea-level rise), causing a double or triple environmental burden for the poor (McGranahan *et al.* 1996). The ecological city approach promotes a cyclical metabolism based on activities such as recycling and reuse of resources and waste.

The business-as-usual and ecological city approaches share two weaknesses: neither adequately synthesises the human and ecological imperatives of sustainable urbanisation, and neither gives adequate weight to the social issues of poverty and vulnerability. The fragmented vision of sustainable urbanisation that these perspectives present allows gains in one area to generate losses elsewhere. A typical example is for wealthy neighbourhoods or cities to export waste, simultaneously addressing local environmental concerns and causing more distant environmental degradation. Sustainable urbanisation requires that urban systems be seen in a regional and global, as well as a local, context and that the components of urban development are viewed as an interacting whole.

The elements that need to be considered in such a holistic view of sustainable urbanisation are shown in Figure 9.3. Five components are identified: social, economic, political, demographic and environmental, and some of the areas in which different sectors interact are shown, though there are many more that could be added. For example, the environment is linked to social development though local environmental quality – the ability of different social groups to access adequate shelter and environmental

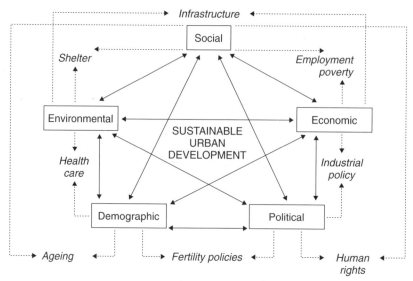

Figure 9.3 Sustainable urbanisation: main components and indicative issues
Source: Drakakis-Smith (1996: v).

infrastructure. This model will be used as a template to discuss contemporary sustainability in a later section of this chapter.

If identifying the components of a more sustainable urbanisation is proving difficult, then operationalisation is no less problematic. Moving towards sustainable urbanisation requires: 'seeking the institutional and regulatory framework in which democratic and accountable city and municipal authorities ensure that the needs of the people within their boundaries are addressed while minimising the transferring of environmental costs to other people or ecosystems or into the future' (Satterthwaite 1997: 1684).

Reorienting cities towards sustainability places emphasis on the need for open and inclusive urban management set within an integrated hierarchy of local, regional and international governance. Inside cities, municipal government occupies a pivotal position in its varied roles of service provider, community resource mobiliser, regulator, advocate and strategic planner. However, the capacity of municipal governments has very often been limited by capital and human resource scarcity, and especially in capital cities, by the capture of municipal responsibilities (and budgets) by central government. Strengthening municipal authorities while broadening the base of urban governance by including civil society and private sector organisations, should be a central concern for policymakers. At the beginning of the twenty-first century, some movement has been made in this direction, with international agencies and international NGOs, as well as some national and city level authorities, prioritising the institutional strengthening of city governments, urban poverty alleviation, family planning and community development. An example of good practice is the rehabilitation of the Rio Torres Grande de Tarcoles, San José, Costa Rica, which involved collaboration between the national and municipal govern- ments and international agencies (Gilbert *et al.* 1996). Region-wide, the Canadian International Development Agency has been particularly active in pressing for good governance, human rights and democracy.

Before assessing the extent to which cities in Middle America have moved towards sustainability it is worthwhile exploring the historical context of contemporary urbanisation in the region. Cities have co-evolved – in the case of Mexico City over three millennia – with local, national and global political, economic, social and environmental forces. The legacy of these forces continues to shape opportunities for and constraints upon sustainability.

Why here, why now? Historicising urbanisation

Dependent urbanisation

Urban form and function in Middle America have been greatly influenced by external political and economic forces. The force of external pressure has led explanations of urbanisation to draw from the school of Depend- ency Theory. Potter (1995) provides a good example of this approach in seeking to explain the distribution and character of urbanisation in the Lesser Antilles.

Potter builds up a three-stage 'Plantopolis' model. In the first stage (mid-1600s to mid-1800s), plantation economics during the era of slavery produced a pattern of self-contained plantation settlements with a single port town emerging as a node for trade and communication with the metropole, local administration and social control. The second stage (mid-1800s to mid-1900s) was stimulated by emancipation and the formation of free villages by people of African descent. The free villages form an inter- mediate tier in the urban hierarchy between the plantation villages and the urban centre. In the modern, post-colonial era (1960 onwards) national development policy and the international division of labour have favoured investment in export-oriented industrialisation and tourism focused upon the principal (capital, primate) port city, and creating an increasingly uneven pattern of urban development (Potter 1995). The Plantopolis model is useful in explaining urban development in the plantation economies of the Eastern Caribbean and to an extent the Guianas, but is less applicable to Central America and the Greater Antilles. Here, external influence was driven by the imperatives of Spanish and French rather than English coloni- alism, and indigenous peoples made a more lasting impact upon European settlement decisions, producing a more varied pattern of urbanisation.

The colonial/mercantile legacy

European colonialism was initially driven by mercantilist trade and the need to accumulate wealth directly from indigenous populations. This favoured interior settlement as a means of maximising trading opportunities.

Suriname and Guyana exemplify this process, where Dutch, French and English merchants established camps up to 100km inland to facilitate easy trade with indigenous peoples (Pelling 1996). In northern Central America and Mexico, the pre-existence of urban cultures at the moment of Spanish conquest also directed colonial settlement away from the coast. The need to control indigenous populations and the symbolic power of pre-colonial cities encouraged Spanish *conquistadors* in Mexico, Honduras, Guatemala and El Salvador to locate on the site of major indigenous settlements in a policy called *congregación*. Thus, Mexico City is built on the ruins of Tenochtitlán, built by the Aztec ruler Montezuma I, who in turn had built on a site first urbanised by the Olmec civilisation in 1300 BC (Lezama 1994).

Competition between colonising powers led to strategic settlement location deviating from purely economic considerations. During the formative stages of Spanish colonisation, towns were often established to support rival territorial claims between competing Spanish factions (for example, the towns of Trujillo and San Jorge Olancho in Honduras) or between competing national powers in border areas, as between the Dominican Republic and Saint Dominigue (Haiti), and Mexico and the US. As the colonial trade between Middle America and Europe grew, military security became a concern, leading to the location of centres away from the coast, such as Villa de la Vega (Spanish Town), the Spanish capital of Jamaica, which lies 10km from the coast (Hill and Bender 1994).

Capitalist urbanisation

Changes in the global political-economy between 1600 and 1950 wrought profound reconfigurations in urban location and function. Most important among these changes have been the end of slavery, the fluctuating profits to be made from primary products, particularly coffee and sugar (see Thorpe and Bennett, this volume), periods of national and international armed conflict, the rise of political oligarchs and natural disasters.

Slavery was abolished in Middle America in the mid-1800s, with Haiti, the world's first free black state, becoming independent from France in 1804 following 14 years of revolt led by Toussaint L'Ouverture. However, in both the free villages of the West Indies and the free nation of Haiti, urban life continued to be dominated by the interests of capital. In its early years of independence, Haiti was a rich and liberally governed nation. However, in 1825 Haiti agreed to pay an indemnity to France in return for national recognition, the huge sums demanded by France left little capital for investment in national services and infrastructure, placing the nation on a path to decline from which it has yet to recover (Tata and Tata 1994). Similarly, in the West Indies, free villagers remained closely tied to the sugar estates for employment. This dominant relationship was forcibly upheld in Guyana by the flooding of free village crop-gardens using sugar estate irrigation waters, causing crops to rot and compelling villagers to seek employment on the estates (Rodney 1981). Emancipation allowed free movement for Afro-Caribbean labour migrants from the West Indies, adding to the ethnic and cultural pluralism of cities in the region. The most important destinations at the end of the nineteenth century were the railway and canal works in Panama, and the banana plantations of Costa Rica (Limón retains an English-speaking minority) and Honduras.

The close ties between economic development and urbanisation are clearly drawn out in the history of Middle American urbanisation. For Central America the coffee boom of the mid-nineteenth century was a period of urban reorganisation and growth (see Thorpe and Bennett, this volume). In Costa Rica, San José, which was founded as a centre for tobacco cultivation in 1737, came to prominence in this period. Costa Rica's second city, Limón, grew most rapidly during two periods of banana plantation cultivation (1883–1927 and 1950 onwards) with an intervening period of decline coinciding with the temporary abandonment of banana cultivation. Guatemala City also benefited from the coffee boom, growing in size and expanding along recently constructed railway lines. Guatemala is a somewhat unusual case in that the liberal government of the time invested some of the large rents acquired from coffee production in infrastructure works to improve the living environment of Guatemala City's expanding slums (Elbow 1994). Sugar was the basis for similar economic growth and urbanisation in Cuba, Jamaica, the Lesser Antilles, Guyana and Suriname in the eighteenth and nineteenth centuries.

Throughout the twentieth century, many parts of Middle America were plagued by armed conflict and civil war. Conflict accelerated urban growth as residents of vulnerable rural areas sought the relative security of the city, and contributed to the deterioration of urban life, both directly as a consequence of military destruction and indirectly as national revenues were diverted away from infrastructure investment. Civil war in El Salvador between 1979 and 1986 displaced 33 per cent of the national population, with most rural refugees moving to safer urban locations (Greenfield 1994). In Panama City, 4,000 houses were destroyed and up to 14,000 people left homeless in a single air-raid by the United States Air Force in 1989, during the forced removal from office of the dictator Manuel Noriega (Greenfield 1994).

National urban planning policy (or the lack of it) in the twentieth century has been shaped by a small number of oligarchs. The rise of the Somoza family in Nicaragua in the 1950s is a prime example. The centre of the Somoza family's interests lay in Managua, where they had large investments in construction, planning and banking. The Somozas were able to exploit the 1972 Managua earthquake, in which 80 per cent of houses were destroyed, by diverting aid through their own business interests (Higgins 1994). Somoza, and similar oligarchs such as Trujillo and Balaguer in the Dominican Republic, pursued processes of partial modernisation, promoting their chosen cities as engines of economic growth and as symbols of national identity and personal power, rather than as sites for providing for the basic needs of residents. The result has been high levels of inequality and political marginalisation (Cela 1996).

The possibility that environmental forces can shape settlement patterns and the quality of urban life is often overlooked. Santiago, the original capital of Guatemala, was abandoned in 1541 following a landslide that swept through the town; the capital was relocated, only to be destroyed by an earthquake in 1733. Jamaica's original capital under British rule,

Port Royal, was destroyed by an earthquake and tidal wave in 1692 and relocated to Kingston. Perhaps the most hazard-prone city is San Salvador, which remains stubbornly at the foot of the active San Salvador volcano despite being destroyed by earthquake nine times between 1575 and 1986. The high earthquake hazard of this region is reflected in the number of low-rise buildings that can still be found in the colonial centres of, for example, San José (Costa Rica) or Santiago de Cuba (Cuba). The El Salvador earthquake on 13 January 2001 once again exposed the vulnerability of urban settlements in this region to natural disaster.

Industrialisation and rapid urbanisation

The most rapid urban growth has taken place since 1950, and the most significant force shaping this has been national industrial development policy. Throughout the region, most state and foreign investment has been directed at the capital city, as an existing centre with a physical infrastructure and appropriate labour force. For many development planners, attracting financial investment to the city and creating an industrial base was critical not only for urban development, but also as a trigger for national economic development. Surplus revenue generated by an industrial city was believed to trickle down to alleviate poverty and to act as a catalyst to expand the national economy and move away from an agricultural production base. This was the case with Puerto Rico's Operation Bootstrap in the 1950s and Barbados' Operation Beehive in the 1960s (Duany 1997). That this occurred when rural employment on coffee and sugar plantations was in decline simply added to the movement of labour from the countryside to the newly industrialising cities, accelerating urban primacy. More recently tourism has been a motor for urban growth, allowing macroeconomic diversification and limited urban decentralisation, for example, Acapulco and Cancún in Mexico.

Where decentralisation policies have been in place, they have often focused on decentralisation within the principal metropolitan zone (as in San José, Costa Rica) rather than attempting national decentralisation or industrial development. However, there are exceptions. In Cuba, urban population growth rates declined from an average of 3 per cent per annum between 1950 and 1970 to 1.7 per cent per annum between 1980 and 1990 (Díaz-Briquets 1994). Havana's slow growth, especially in the 1970s, reflected government policies promoting growth in regional centres, but also mass emigration to the US, a declining national fertility rate and the capacity of the government to place strict controls on the allocation of housing and employment. In the Anglophone Caribbean, high emigration rates to the UK, US and Canada have slowed the growth of urban centres. In Guyana, political and economic crisis in the 1970s and 1980s stimulated emigration rates exceeding birth rates, and resulted in negative national and urban population growth rates. Emigration has also been encouraged in the French Caribbean as a mechanism for population control, yet this has faltered in the 1980s and 1990s in response to changing economic conditions in metropolitan France and increasing return migration (Condon and Ogden 1997).

Population growth rates in Middle America's largest urban agglomeration, Mexico City, have been declining, with growth rates of less than 1 per cent per annum, compared to 5 per cent per annum between 1950 and 1979. Slower urbanisation is associated with dramatic falls in natural population growth rates, partly linked to economic recession in the 1980s. People were also less inclined to migrate to Mexico City and even returned to the countryside as the economic advantage of an urban base was seen to deteriorate. Economic restructuring led to the flight of manufacturing companies from central Mexico City as they sought to avoid traffic congestion and high land costs by relocating to nearby urban centres. NAFTA further shifted industrial advantage from Mexico City to the US border towns, stimulating investment and population growth in Tijuana, Mexicali and Ciudad Juárez (Gilbert 1995; Oliveira and Roberts 1996).

This historical review has demonstrated the ways in which urbanisation in post-Columbian Middle America has been characterised by dependency, diversity and change. Dependency on global markets has produced boom and bust economies and consequent periods of rapid urban growth and decline. Diversity between and within cities reflects ethnic, religious and political plurality, and demonstrates that locality matters in shaping peoples' experience of urban life. But perhaps above all, the region's urban centres and their inhabitants have shown a great dynamism. The richness of the region's urban life lies in the struggles and victories of everyday life in a turbulent world.

Challenges to sustainable urbanisation in Middle America

This section is organised around the five components identified in Figure 9.3 (page 224). The division of urban life into separate components is an over-simplification of reality, but it allows a focused discussion on contemporary urban problems. The status of each component and its contribution to sustainability is explained and illustrated by an example from the region.

Environment

A degraded environment is symptomatic of exploitative relations between different groups in society as well as between society and nature. In almost all large urban centres, planning weaknesses, poverty and rapid urban population growth have led to the exclusion of considerable numbers of people from the formal housing market and environmental services. The growth of inner-city slums and expansive fringe squatter settlements is the result, leading to increased health risks for their residents. Most losses to hurricane 'George' in Santo Domingo, 1998, were in fringe squatter settlements. Local environmental risks are often compounded by the development externalities of urban industrial and transport infrastructure. In Mexico City since the early 1990s, air pollution has been an almost constant health risk: during 1997 levels of ozone were 1.7 times the acceptable limit on 90 per cent of days (Connolly 1999). As cities become larger, more

---

Box 9.1   Environmental restoration in Havana

Urban agriculture reduces the transport costs of providing food, increases the opportunity for urban dwellers to eat a balanced diet and creates productive green spaces in the city. Havana's system of *organopónicos* (organically farmed urban gardens) was initiated in 1987 as part of a national food security programme. Collective groups turn unproductive open space into urban market gardens. Vegetables are grown intensively but organically (because of limited access to chemical inputs as well as ecological thinking) and sold to local consumers. In 1998 Havana's *organopónicos* produced over one million tonnes of vegetables (see Plate 9.1).

In 1981 the Office of the Historian of the City of Havana was awarded funds from central government to restore the architectural heritage of the city. In 1982, Havana's colonial quarter was declared a World Heritage Site by UNESCO. By 1999 the Office of the Historian had completed over 200 restoration works. Success has come from a circular budget where a proportion of restored buildings have been given to tourist use as hotels or restaurants. The rent from these businesses is returned to the Office of the Historian and reinvested in restoration. Buildings of architectural importance have been saved from decay but the social benefits are unclear. The residents of buildings given to tourist uses have been forced to relocate, however, on the other hand, projects targeting low-income neighbourhoods for restoration have improved the quality of life for residents that remain.

Sources: Granma (1999); Office of the Historian (1999).

---

industrialised and wealthier they have a growing impact on ecology outside the built-up area. Mexico City is the largest urban area in the region and its demand for water alone extends an 'ecological footprint' into hydraulic systems in neighbouring valleys up to 150km distant.

Promoting a positive urban environment and sustainable consumption and waste production practices are low on the agenda of most cities in the midst of immediate economic and political crises. However, there are opportunities for environmental goals to be fused with economic and social development. Box 9.1 presents the example of Havana, Cuba, and examines the city's success in supporting urban agriculture and restoring the city's built heritage.

Demographic

Population status or demographic change affects urban sustainability in a number of ways. It can influence the size of the active population relative to those too young or old to work, the size of the urban labour force relative to the job market, demand for services such as education and healthcare, and social composition and the likelihood of social unrest. The size and structure of urban populations are influenced by migration (movement

Plate 9.1   Urban agriculture: organically farmed urban gardens, Barrio
Cubanacán, Havana, Cuba
Photo: Mark Pelling

in and out of the city, as well as from one part of the city to another) and
natural change (the change in a population's size accounted for by births
and deaths).

As cities grow, in-migration tends to be replaced by natural change as
the principal motor of urban population expansion. In Middle America,
rural poverty combined with the coffee and sugar booms of the mid- to
late 1800s and industrialisation after 1950 to form two waves of migration-
led urban growth. However, since the late 1980s migration has been over-
taken by natural growth as the principal component of urban population
increases, and family planning has become an important component of
urban population control programmes. Box 9.2 examines some of the environmental implications of changing population patterns in Mexico City.

Sustainable cities require healthy populations. Health care in Middle
American cities has improved over the last 30 years through the implementation of low-cost basic drug and primary health care provision. However, the imposition of Structural Adjustment Programmes (SAPs) in the
1980s and 1990s reduced public spending on health, creating widening
inequalities in access to care between and within cities (see McIlwaine,
this volume). Only Nicaragua and Cuba maintain national health systems
run through the public sector, with most countries having a hybrid system
of public/private care (Potter and Lloyd Evans 1998). In all cities across
the region ill-health continues to be associated with inadequate living and
working environments (see above), producing morbidity and mortality

231

Box 9.2 Population and the environment in Mexico City

Despite predictions to the contrary, Mexico City's population appears to have stabilised in the 1990s, with in-migration no longer being seen as a major issue in urban development. Between 1987 and 1992 there was a net negative in-migration of 180,000 for the whole metropolitan region. Overall annual growth rates for Mexico City support this claim: in 1950 the growth rate was 6.7 per cent, in 1970 it was 5.6 per cent, in 1990 it had decreased to 2.3 per cent with a further decline in 1995 to 1.9 per cent.

Lower growth rates can be explained by a national financial crisis since 1982 and the de-industrialisation of Mexico City in the 1990s, with subsequent reductions in formal sector employment. The reduction of Mexico City's advantage in socio-economic and quality-of-life indicators relative to other smaller urban centres over this period also explains why in-migration flows have slowed. Wealth-related environmental problems have been important in declining quality-of-life indices with traffic congestion and atmospheric pollution reaching critical levels in the 1980s.

Population has responded to environmental degradation, but not all population groups have responded in the same way, and this has contributed to a new urban geography in Mexico City. Migrants into the city are characterised by higher levels of poverty and lower levels of education than their counterparts of 20 years ago. Migrants leaving the city tend to be better qualified and are moving to peripheral urban developments where environmental problems can be avoided, contributing to urban sprawl. Sprawl has been compounded by the delayed effects of the city's demographic transition. The reduction of the fertility rate from five to two between 1970 and 1995 decreased household size but has still led to an increased demand for housing. The inequalities brought about by migration and urban sprawl are exacerbated by Mexico City's system of urban government, which is decentralised into a number of municipalities, each raising its own budget depending upon its financial base. Consequently the poorer municipalities that are receiving most of Mexico City's low-income population growth are unable to invest in physical or social infrastructure, with the effect that, despite an overall decline in population growth, inequality in access to environmental goods and a safe living environment continues to rise.

Source: Connolly (1999).

---

from preventable diseases such as cholera, malaria, yellow fever, dengue fever and intestinal parasites. Since the 1980s, AIDS has become a growing concern, and one that is often suppressed for fear of losing tourist revenue. HIV / AIDS appears to be most common in the Caribbean, with 1.8 per cent of the subregion's population having tested positive. The highest rates are found in Haiti, where 5.2 per cent of the population has tested positive for HIV / AIDS (PAHO 1998).

Middle America is an important source for international migration to North America and Europe, and for some cities, such as Georgetown, Guyana, emigration rates have been sufficiently high to suppress urban population growth. The Caribbean diaspora has led to the formation of transnational families, facilitating cultural and economic exchange between the Caribbean and, in particular, North America. The remittance of money or goods from migrants to family members remaining in the Caribbean has come to play an important survival strategy and helped to sustain many poor urban households (see also Willis and McIlwaine, this volume).

Social

Middle American cities are noted for their ethnic and religious diversity. For example, Paramaribo, Suriname, has six major ethnic groups – Creoles of African descent, East Indians, Javanese, Chinese, Amerindians and Europeans, and adherents of Bahá'í, Christianity, Hinduism, Islam and indigenous religions. Diversity is a seedbed for cultural hybridity and the formation of distinctive Caribbean and Central American identities, but is also a source of social tension. Ethnic social division is too common and has resulted in inequalities in access to political and economic power (Gilbert 1990; Brown 1999). In Central America the *blancos* or *mestizos* with European heritage have often dominated people with indigenous ancestry or cultural affiliation. In the Caribbean Islands post-colonial societies have struggled to overcome racial discrimination (see Howard, this volume).

Increasing urban growth and poverty have undermined social connectivity in many poor urban communities. This is especially so where growing crime and violence have been fuelled by the drugs trade. Violence can undermine intra-household relations, break down trust and co-operation between members of the same community, and encourage the formation of new social organisations such as gangs, illegal organisations or protection groups (Moser and McIlwaine 2001). In Box 9.3, the work of one NGO to help rebuild social capital in a marginal *barrio* in Santo Domingo, the Dominican Republic, is discussed (see also McIlwaine, this volume, on definitions of social capital).

Political

One of the central tenets of sustainable urbanisation is the need for decision-making to be inclusive and transparent, and for decision-makers to be accountable to stakeholders affected by urbanisation – including those who may not reside in the city. There has been some success in integrating different urban political actors in a move towards urban governance that incorporates civil society and private-sector organisations with the public sector in urban decision-making. Including the voices of more distant stakeholders (in time and space) is proving more difficult. Fiszbein and Lowden (1999) provide several examples of good practice in this region where international private sector, local government and community actors have worked together. Cases include a primary school project in Ilpango, El Salvador, and COMAND – a low-cost housing programme in

---

Box 9.3   Los Manguitos, Santo Domingo,
the Dominican Republic: building social capital

Los Manguitos grew in the 1970s as rural migrants flooded Santo Domingo. However, since the mid-1990s an increasing number of residents have sought to move out of the *barrio* (settlement) because of rising levels of drug-related crime and violence. Directly confronting organised drug crime is beyond the capacity of any single NGO, but one organisation, the *Instituto Dominicano Desarrollo Integral* (the Dominican Institute for Integrated Development – IDDI) has succeeded in strengthening social cohesion in Los Manguitos and allowed people to confront their own experiences of violence – domestic and public. Plate 9.2 shows part of a women's group who attend weekly meetings at IDDI's office in Los Manguitos and who also work as community health promoters.

Social capital has been built through the women's group, which has 40 members, but also through IDDI's work with local community-based organisations. There is a history of politicisation of community organisation and a resultant lack of co-operation, but IDDI has been able to bring six groups together through the formation of a jointly managed neighbourhood rubbish collection business. This too has strengthened the community's social capital and provides an alternative to social support networks based on political patronage or the illegal drugs business.

Source: Pelling (2000).

Jamaica. Inclusivity is hampered by a legacy of authoritarian city leadership, modelled upon national leadership styles, which has suppressed civil society to varying degrees. External influence is always at hand, initially through the European colonial powers, then through the shifting repercussions of the Cold War and most recently through the growing reach of neoliberal ideology and international financial institutions. The power of external actors to shape urban policy is shown most vividly through the IMF-sponsored SAPs, enacted across the region in the 1980s and 1990s, which were associated with a deepening of poverty (see McIlwaine, this volume), a reduction in the state provision of basic services, and an increased role for the private sector in urban management. The relationship between city and central government has been critical for urban policy formation. In Santo Domingo, most of the municipality's responsibilities (and accompanying budget) have been captured by the central government, weakening the city's influence. In the small cities of the Caribbean, the local or municipal tier of government is absent, with any decision-making taking place in the national parliament. The example of Georgetown, Guyana, in Box 9.4 highlights the difficulties faced in reforming urban political systems.

Plate 9.2 Women's group, Los Manguitos, Santo Domingo, Dominican Republic
Photo: Mark Pelling

Economic

Economic status shapes the distribution of assets within society and the opportunities for individuals and city managers to achieve quality of life without degrading the environment. The late 1970s, 1980s and 1990s were a period of economic crisis for much of urban Middle America; for example, unemployment in Panama City reached 21 per cent in 1988 and Mexico City lost close to a quarter of its manufacturing jobs in the 1980s. Reduced formal sector employment resulted in expansion of the informal sector. In Mexico City:

> swarms of young jugglers, windscreen washers and fire-eaters appeared at major road junctions. The metro was suddenly awash with traders; every carriage had its vendor of pens, razors, maps or magazines; when no one was selling goods people were singing for their supper.
>
> (Gilbert 1995: 327).

The capacity of the informal sector was soon overwhelmed as the number of workers grew and poverty deepened (see McIlwaine *et al.*, this volume). Formal sector workers fared little better: in Mexico City formal sector wages had fallen to 76 per cent of their 1980 value by 1990 (Gilbert 1995). Increasing urban poverty has meant that more people are excluded from the formal housing market and that urban authorities have been increasingly unable to invest in public housing. Informal settlements have grown

Box 9.4   Negotiating urban politics in Georgetown, Guyana

Georgetown, the colonial 'Garden City of the Caribbean', has been reinvented by the local media as the 'Garbage City of the Caribbean' and is plagued by environmental problems such as uncollected household rubbish, blocked drains and flooding. The Georgetown City Council receives barely enough revenue from rates (only 66 per cent paid in 1992) to pay staff wages, let alone provide basic services to the city. Government grants are available but have often been withheld because of a lack of financial transparency in City Hall. Political tensions between the city and national government are perhaps a deeper cause. The two extracts below are taken from newspaper articles and they sum up residents' frustration at the deleterious effect that political competition has had on the urban environment.

> The most serious problem seems to be that the war of attrition between the People's Progressive Party (PPP-Civic) administration in the Central Government and the For a Good and Green Guyana (GGG) party that dominate City Council has finally eroded the basis for co-operation between the two. . . . [I]nitiatives have been taken by the Central Government without consultation with the City Council; a political move designed to embarrass the latter but which will diminish the effectiveness of the entire programme . . . A physical solution to flooding could be provided but the political impasse between the PPP-Civic and GGG impedes progress. Is the capital being punished for electing the wrong mayor?
>
> *(Guyana Review* 1997).

> What I can't tolerate any more is the 'stinking' City Council. I cannot accept that its elected Council members should continue spending time scoring cheap political points off each other while Georgetown dies.
>
> (Anadaiye 1996).

Sources: Anadaiye (1996); *Guyana Review* (1997).

---

as a result, with large numbers of people being excluded from adequate healthcare, education or environmental infrastructure.

As poverty has continued, gendered labour divisions have changed. More women are now engaged in the workforce. In Georgetown, Guyana, labour force participation rates for women rose from 28 per cent in 1970 to 47 per cent in 1992, with the largest increases in women between the ages of 25 and 39 (Peake and Trotz 1999). Despite women's participation in the labour force, their wages are usually lower than men's and they also have to cope with domestic duties (see Chant, this volume). With economic recession children have also become active in urban workforces. Box 9.5 shows the efforts made by one community-based organisation to combat poverty in Bridgetown Barbados.

Box 9.5 Poverty alleviation in Bridgetown, Barbados

The community-based organisation (CBO) Pinelands Creative Workshop (PCW) was established in 1975 in the community of the Pine, a low-income government housing scheme on the outskirts of Bridgetown. The community has faced a number of barriers to economic development since the mid-1970s and at each stage PCW has been able to respond by redirecting its anti-poverty programme. Initially the Pine had a bad reputation, which made employers reluctant to hire local workers, and was famed for the 'three Ps': 'If you were from the Pine and you came before Magistrate Perry's court you were sure to be sent to Prison.' Creating a more positive image for the residents of the Pine was the first response of PCW, and this was achieved over a long-term project based on cultural workshops and performances. Today PCW has a professional dance company that performs to tourists in the City Museum and on cruise ships, the profits of which sustain the organisation's administrative needs.

In the 1990s, more direct action has been taken to confront poverty. PCW has been able to mobilise the community through organising adult literacy and work skills programmes, providing grants for education to seven students from low-income households, job placements for young unemployed community members and a meals-on-wheels service to 77 senior citizens. In 1998 PCW was sponsored by the Government of Barbados to run a 'sustainable development in low-income communities programme'. A trust fund has been established and PCW awarded B$188,000 (approximately US$94,000) to 94 local small businesses. The story of one recipient is below:

> Mad Dog is 30 years old and lives alone in the Pine. He received a grant of B$10,000 (approximately US$5,000) in 1998, as well as business management training that enabled him to open a hot food and snack bar called 'Wholesome Foods' in the local area. Before this he was employed as a plumber and worked part-time selling cooked food at weekends.
>
> Mad Dog first heard about the programme through a friend. He has made all the repayments on his loan and employs four people part-time. The business is well managed and the food product is selling. The major problem facing the business is that customers who previously drove in from outside the community have been deterred by increasing levels of gun violence in the Pine.

Source: Pelling (1998).

Conclusion

The question of what form sustainable urbanisation might take in Middle America was posed in the introduction to this chapter. There is no doubt that urbanisation has been an enduring aspect of human organisation in the region, yet its sustainability is less assured. Since European colonisation, the greatest challenge to sustainability has been cities' dependency upon external political economies. When cities have not been able to channel local surplus product into the external (colonial/global) market place, they have declined; when cities have succeeded they have been the sites of rapid population growth and unsustainable environmental and social relations.

Drakakis-Smith's (1996) model allowed us to unpack the meaning of sustainable urbanisation and to focus on the various sectors that together make up urban space. In doing so the interaction between the five elements of sustainable urbanisation was clear to see. But, in the present period of rapid urban growth, can this help us move closer to identifying how a more positive urban future may come about? The examples above suggested that positive outcomes are possible when individuals, community organisations and governments work together to confront poverty, marginalisation and environmental degradation. Co-operation provides a mechanism for the resource attributes of each institutional level to be brought together and for building representativeness and responsiveness in urban management. However, as Mad Dog (Box 9.5) has been finding out in Bridgetown, whilst working together for change can be effective, the future is always uncertain and actors need to be increasingly aware of their place in the shifting web of urban (and global) life to succeed. Policy actions have reactions, and it is the impact of these reactions, which increasingly jump across sectoral boundaries and have a distanced impact in time and space, that needs to be given increased weight if a more sustainable urbanisation in Middle America is to be realised.

Useful websites

For detailed data on individual cities it is also worth consulting national websites. International organisations with interests in urbanisation worldwide also hold relevant information. The sites below are of this second type, and are all attached to UN organisations.

www.unfpa.org/regions/lac/index.html United Nations Population Fund, Latin America and the Caribbean site.

www.unchs.org/unchs/english/hagenda United Nations Centre for Human Settlements, Habitat II agenda.

www.unchs.org/scp/ United Nations Centre for Human Settlements, Sustainable Cities Programme.

www.bestpractices.org/ Database of best practices for human settlements.

Further reading

Drakakis-Smith, D. (1995) Third World cities: sustainable urban development I, *Urban Studies* **32**(4–5): 659–677.

Drakakis-Smith, D. (1996) Third World cities: sustainable urban development II, *Urban Studies* **33**(4–5): 673–701.

Drakakis-Smith, D. (1997) Third World cities: sustainable urban development III, *Urban Studies* **34**(5–6): 797–823.
This three-paper series provides an excellent overview of the concepts and praxis of sustainable urbanisation. The first paper sets out the five-component model of sustainable urbanisation used in this chapter, with the various components being reviewed in this and the subsequent papers.

Greenfield, G.M. (ed.) (1994) *Latin American Urbanisation: Historical Profiles of Major Cities.* Greenwood Press, London.
Provides a useful overview of urbanisation processes in Latin America, including Mexico and Central America.

Potter, R.B. (2000) *The Urban Caribbean in an Era of Global Change.* Ashgate, Aldershot.

Potter, R.B. and **Conway, D.** (1997) *Self-Help Housing, the Poor and the State in the Caribbean.* University of the West Indies Press, Jamaica.
Both volumes provide excellent starting points for a consideration of urbanisation in the insular Caribbean.

Pugh, C. (ed.) (2000) *Sustainable Cities in Developing Countries.* Earthscan, London.
Presents a multidisciplinary range of innovative thinking on sustainable urbanisation.

Satterthwaite, D. (ed.) (1999) *Earthscan Reader in Sustainable Cities.* Earthscan, London.
Brings together a huge range of papers by experts in the theoretical and practical aspects of sustainable development as it might be applied to cities.

References

Anadaiye (1996) Of a dead dog and a stinking council, *Stabroek News,* 26 June 1996.

Brown, D. (1999) Ethnic politics and public sector management in Trinidad and Guyana, *Public Administration and Development* **19**(4): 367–379.

Cela, J. (1996) La ciudad del futuro o el futuro de la ciudad, *Antología Urbana,* Ciudad Alternativa, the Dominican Republic.

Condon, S.A. and **Ogden, P.E.** (1997) Housing and the state in the French Caribbean, in Potter, R.B. and Conway, D. (eds) *Self-Help Housing, the Poor and the State in the Caribbean.* University of the West Indies Press, Mona, Kingston, Jamaica: 217–242.

Connolly, P. (1999) Mexico City: our common future? *Environment and Urbanization* **11**(1): 53–78.

Díaz-Briquets, S. (1994) Cuba, in Greenfield, G.M. (ed.) *Latin American Urbanisation: Historical Profiles of Major Cities*. Greenwood Press, London: 173–187.

Drakakis-Smith, D. (1996) Sustainability, urbanisation and development, *Third World Planning Review* **18**(4): iii–x.

Duany, J. (1997) From the Bohío to the Casrío: Urban housing conditions in Puerto Rico, in Potter, R.B. and Conway, D. (eds) (1997) *Self-Help Housing, the Poor and the State in the Caribbean*. University of the West Indies Press, Mona, Kingston, Jamaica: 188–216.

Elbow, G.S. (1994) Costa Rica, in Greenfield, G.M. (ed.) *Latin American Urbanisation: Historical Profiles of Major Cities*. Greenwood Press, London: 159–172.

Fiszbein, A. and **Lowden, P.** (1999) *Working Together for a Change: Government, Civic, and Business Partnerships for Poverty Reduction in Latin America and the Caribbean*. World Bank, Washington, D.C.

Gilbert, A. (1990) *Latin America*. Routledge, London.

—— (1995) Debt, poverty and the Latin American City, *Geography* **80**(4): 323–333.

Gilbert, R., Stevenson, D., Girardet, H. and **Stren, R.** (1996) *Making Cities Work: the Role of Local Authorities in the Urban Environment*. London, Earthscan.

Girardet, H. (1999) *Creating Sustainable Cities*. Schumacher Briefings No. 2, The Schumacher Society, Bristol.

Granma (1999) Una práctica agroecológica adecuada a las nuevas condiciones sociales y financieras, 14 January.

Greenfield, G.M. (ed.) (1994) *Latin American Urbanisation: Historical Profiles of Major Cities*. Greenwood Press, London.

Guyana Review (1997) Capital Punishment, January.

Hardoy, J., Mitlin, D. and **Satterthwaite, D.** (1992) *Environmental Problems in Third World Cities*. Earthscan, London.

Hill, N.P. and **Bender S.A.** (1994) Jamaica, in Greenfield, G.M. (ed.) *Latin American Urbanisation: Historical Profiles of Major Cities*. Greenwood Press, London: 396–415.

Higgins, B. (1994) Nicaragua, in Greenfield, G.M. (ed.) *Latin American Urbanisation: Historical Profiles of Major Cities*. Greenwood Press, London: 396–415.

Lezama, J.L. (1994) Mexico, in Greenfield, G.M. (ed.) *Latin American Urbanisation: Historical Profiles of Major Cities*. Greenwood Press, London: 350–395.

McGranahan, G., Songsore, J. and **Kjellén, M.** (1996) Sustainability, poverty and urban environmental transitions, in Pugh, C. (ed.) *Sustainability, the Environment and Urbanisation*. Earthscan, London: 103–134.

Moser, C. and **McIlwaine, C.** (2001) *Violence in a Post-conflict Context: Urban Poor Perceptions from Guatemala*. World Bank, Washington, D.C.

Office of the Historian (1999) *Desafío de una utopía*. Oficina del Historiador, La Habana.

Oliveira, O. de and **Roberts, B.** (1996) Urban development and social inequality in Latin America, in Gugler, J. (ed.) *The Urban Transformation of the Developing World*. Oxford University Press, Oxford: 253–314.

Pan-American Health Organisation (PAHO) (1998) *AIDS Surveillance in the Americas* (available on PAHO website *http://165.158.1.110/english/aid*).

Peake, L. and **Trotz, D.A.** (1999) *Gender, Ethnicity and Place: Women and Identities in Guyana*. Routledge, London.

Pelling, M. (1996) Coastal flood hazard in Guyana: environmental and economic causes, *Caribbean Geography* **7**(1): 3–22.

—— (1998) *Sustainable Development in Low-Income Communities*. Internal report: Pinelands Community Workshop, Bridgetown, Barbados.

—— (1999) The political ecology of flood hazard in urban Guyana, *Geoforum* **30**: 249–261.

—— (2000) *Santo Domingo* working paper No. 4, Social Capital, Sustainability and Natural Hazards in Caribbean Microstates project, Department of Geography, University of Liverpool.

Potter, R.B. (1995) Urbanisation and development in the Caribbean, *Geography* **80**(4): 334–341.

Potter, R.B. and **Lloyd Evans, S.** (1998) *The City in the Developing World*. Longman, Harlow.

Rodney, W. (1981) *A History of the Guyanese Working People, 1881–1905*. The Johns Hopkins University Press, Baltimore and London.

Satterthwaite, D. (1997) Sustainable cities or cities that contribute to sustainable development? *Urban Studies* **34**(10): 1167–1691.

Tata, R.J. and **Tata, S.J.** (1994) Haiti, in Greenfield, G.M. (ed.) *Latin American Urbanisation: Historical Profiles of Major Cities*. Greenwood Press, London: 294–312.

UNFPA (United Nations Population Fund) (1999) *The State of World Population, 1999* (available at *www.unfpa.org/swp/1999*).

United Nations Development Programme (UNDP) (1999) *Human Development Report 1999*. Oxford University Press, New York and Oxford.

WCED (World Commission on Environment and Development) (1987) *Our Common Future*. Oxford University Press, New York and Oxford.

World Bank (2000) *World Development Report 1999/2000*. Oxford University Press, New York and Oxford.

CHAPTER 10

Social roles and spatial relations of NGOs and civil society: participation and effectiveness post-hurricane 'Mitch'[1]

Sarah Bradshaw, Brian Linneker and Rebeca Zúniga

Introduction

In Central America, the 1980s were characterised by political negotiations in search of peace, while the 1990s witnessed initiatives aimed at the consolidation of emerging democracies and governability. Non-governmental organisations (NGOs) are playing an important role in the consolidation of democracy through the development of organised civil society, along with the emergence of popular social forces and broader citizen participation in the decision-making process (Fundación Arias 1997; Serbin 1998). In Nicaragua, Honduras, El Salvador and Guatemala in particular, these emergent social forces, by changing the traditional alliances between the agricultural oligarchy, the military and external forces, are seen to be ways of consolidating democracy and avoiding the return to 'reactionary despots' of the past (Karl 1995).

This chapter considers the developing social roles and spatial relations of civil society and, more specifically, of NGOs in Middle America. It considers their effectiveness in terms of both their traditional welfare roles as service providers, and in influencing the socio-economic and political development of the region through their policy advocacy roles. It focuses geographically on the Central American region, and more specifically, on the evolution and characteristics of civil society in the wake of hurricane 'Mitch'. The first section considers some conceptual and theoretical perspectives on civil society, social movements, along with their evolving social roles in relation to donors and beneficiaries, and the nature of their policy advocacy roles. The second section examines the developing spatial

[1] The views expressed in this chapter are those of the authors as audience and narrators, not as actors, and do not represent the views of the institutions for which they work.

243

network structures of civil society and NGOs at national and regional levels. This also includes consideration of some recent regional-level initiatives in relation to policy decision-making and advocacy. The final section presents a case study of the Civil Co-ordinator for Emergency and Reconstruction (*Coordinadora Civil para la Emergencia y la Reconstrucción – CCER*) in Nicaragua – a civil society co-ordinating organisation. It analyses the involvement of CCER in the reconstruction and transformation process post-'Mitch' in relation to the state and the international community. It explores how the sector has provided relief and reconstruction services, as well as funding issues and the progress and pitfalls of state–civil society relationships.

Over the last 20 years, the effects of globalisation, neoliberal structural adjustment policies (SAPs), along with the differential impacts of regional integration policies in wider Middle America, have left many nations exposed to socio-economic, political and physical vulnerabilities. These have been particularly pronounced in the poorest Central American countries of Nicaragua, Honduras, El Salvador and Guatemala. Their vulnerabilities were recently starkly highlighted when hurricane 'Mitch' hit Central America in October 1998, preceded by hurricane 'George' in the Caribbean earlier in the same year. In Central America, these factors provoked one of the worst disasters in over 200 years (CINDI 1999). Hurricane 'Mitch' affected almost 3.5 million people, with 18,000 dead or disappeared. Overall, losses in the region were estimated at over $US6 billion, with Honduras and Nicaragua experiencing the worst impacts (CEPAL 1999). In Nicaragua, more than 870,000 people were affected, and over 3,000 died (Linneker *et al.* 1998). A large proportion of the population lost their homes, land and means of survival, with the poorest being the most vulnerable as they usually resided in the areas most exposed to flooding and landslides.

However, as well as widespread destruction, hurricane 'Mitch' also acted as a catalyst for the organisation of civil society in Central America. There were hopes that the reconstruction efforts would create new links between civil society, national and local governments, and the international donor community in constructing sustainable development strategies which focus on people, and in particular, the poor and marginalised sectors of society. As well as outlining the nature of civil society in Middle America more generally, this chapter explores the extent to which these processes have actually occurred with particular reference to Nicaragua.

Conceptual perspectives on civil society, NGOs and social movements

Both NGOs and social movements are important constituent parts of civil society. This section outlines definitions of all three categories and terms from a conceptual perspective, before moving on to explore how civil society in Central America has responded in the aftermath of hurricane 'Mitch'.

Civil society

It has been suggested that there are five main arenas within the social space of a modern consolidated democracy that have implications for governability. These are composed of economic society, civil society, political society, the state apparatus and the judicial system (Linz and Stepan 1996). The term 'civil society' generally refers to any organisation that mediates between the individual and the state, based on a right to associate (Fundación Arias 1998). This rather vague definition of civil society highlights a key problem in that there is a wide range of different definitions and interpretations (see McIlwaine 1998a, 1998b on these debates). Linz and Stepan (1996: 7) define civil society as:

> that arena of the polity where self-organising groups, movements and individuals, relatively autonomous from the state, attempt to articulate values, create associations and solidarities, and advance their interests. Civil society can include manifold social movements (women's groups, neighbourhood associations, religious groupings, and intellectual organisations) and civil associations from all social strata (such as trade unions, entrepreneurial groups, journalists, or lawyers).

As Linz and Stepan suggest, it is important to note the heterogeneity of civil society. This arises not least since civil society has many different expressions, spaces and actors. Thus, conflicts and disagreements often characterise civil society. Indeed, it is important to challenge the common portrayal of civil society as a coherent and homogeneous sector. Many civil society organisations are more formalised and/or recognised than others, which results in differential levels of power. The term 'organised civil society' (OCS) may thus be more useful as it puts the focus on actors such as unions, social movements and NGOs.

As well as variations in definitions, there are also a series of theoretical approaches in understanding civil society. At the outset, it should be stressed that civil society as a 'sphere of social reproduction' has always existed and enjoys a long history of discussion by thinkers such as Hegel, Marx and Gramsci (see McIlwaine 1998b). However, prompted mainly by democratisation processes in the former Soviet Union, as well as in Middle and Latin America, the idea of 'civil society' has re-emerged as a key phenomenon of contemporary development. There are three main theoretical approaches. First, from the neo-conservative perspective, civil society organisations are seen as private sector actors, a view which some suggest advocates the privatisation of both development and democracy. Therefore, this perspective is most closely linked with neoliberal SAPs involving state withdrawal and privatisation (Toye 1987). The neo-conservative approach is also important within the recent 'New Policy Agenda' framework fostered by the international community. Within this agenda, international donors view NGOs as key representatives of civil society, through which they can work to strengthen civil society (albeit for their own ends). The encouragement of NGO activities is economically and politically expedient for donors; NGOs can be used as alternative service providers and as evidence of notions of deepening democracy and good governance. For many NGOs, civil society organisations and social movements, this approach can generate contradictions and problems, especially in donor–NGO relations, and civil society–state relations, linked to issues such as representability, legitimacy and accountability (Edwards and Hulme 1996).

The second approach, known as the liberal pluralist perspective, views the organisations of civil society as focal points for individual political participation and as a way of countering the power of authoritarian states. Therefore, civil society acts as a catalyst for social movements and as a challenge to the concentration of power in state hands. The liberal pluralist perspective is most commonly associated with Northern and Southern NGOs not linked with (and often highly critical of) international financial institutions such as the World Bank.

The post-Marxist approach sees civil society organisations and the state as intertwined. In addition, they highlight the need to recognise the internal conflicts within civil society and the essential role of power relations in governing society. In particular, the post-Marxists have been important in allowing issues of power dynamics in households and women's participation in civil society to be addressed (MacDonald 1997 on Central America; McIlwaine 1998a on El Salvador).

Another related issue linked with the post-Marxist viewpoint is that strengthening civil society can also have negative outcomes. Social relations between the government and civil society can sometimes be conflictive, representing what Foley and Edwards (1996: 142) call the 'paradox of civil society'. Here, the positive democratic effects of civil society's role as a series of pressure groups are contrasted with a potentially damaging situation that can arise when a strong politically independent civil society puts excessive demands on the government. The latter may not be consistent with democracy or governability; in other words, it may involve force or violence.

Social movements

Globalisation and democratisation processes in the region have given rise to the development of 'new' social movements. While the older social movements organised themselves around class issues, the newer social movements articulate themselves around broader societal issues, such as gender, lifestyles, racial inequality and conflict, as well as the environment. New social movement organisations tend to politicise previously non-politicised spaces and connect the local with the global by linking their activities to grass-roots organisations, together with national and international NGOs (MacDonald 1994 on Central America). In order to promote their interests they often use mass mobilisation as a form of pressure to defend or change existing society. Also important to point out is that the distinction between social movements and the second important component of organised civil society, NGOs, is often confused, especially as NGOs have increasingly addressed more 'political' objectives. While in the 1980s, social movements played a key role in the democratisation processes in the region, in the 1990s NGOs have risen to the fore as key social actors. Given this importance of NGOs, the rest of this section will focus on NGOs rather than social movements within civil society.

What are non-governmental organisations?

The upsurge of NGOs as key players in development processes, sometimes referred to as the 'third sector' (beyond the state and private sector), is one of the distinctive features of the last few decades in Middle America and beyond. The initiatives of multilateral and bilateral organisations to formalise and 'foster' organised civil society, and the changing socio-political climate of the region allowing the endogenous development of NGOs, have both played a role. The rise in importance of NGOs can be seen not least in terms of the amount of money channelled through them for development projects every year. OECD figures (1995) (cited in Scott and Hopkins, 1999) suggest that official contributions to NGOs from bilateral assistance, negligible in the 1980s, rose to US$1.04 billion in 1994. Smillie and Helmich (quoted in Scott and Hopkins, 1999) estimate that, globally, US$10 billion annually is channelled through the NGO sector.

The term 'non-governmental organisation' came into being with the passing of Resolution 288 by the UN Economic and Social Council in 1950, defined as being an organisation with no governmental affiliation with consultative status with the UN. The original meaning of the term NGO, however, has been lost and instead it has become a catch-all phrase used to describe any organisation that pertains to civil society and is not directly dependent on the government. The debate around the 'correct' usage of the term NGO has grown in recent years, alongside the production of ever more detailed classification systems (see Vakil 1997 for a useful summary of the debates). Confusion over what constitutes an NGO is largely due to its negative definition in terms of what it is not, rather than what it is. The following three issues provide a useful guideline for defining an NGO:

- *Independence*: they are not dependent on political processes and are independent from national governments.
- *Operation*: they do not seek to maximise profits and do not distribute earnings to the individuals who exercise control within the organisation. Instead, their actions are based on some idea of non-profit, human solidarity or voluntarism.
- *Focus*: they work on development assistance, disaster relief or human rights in developing countries, either directly through working with local people or indirectly through advocacy (the work of raising people's awareness of issues).

Within these broad guidelines are many variations used to describe different types of NGOs. These include BINGOs (big international non-governmental organisations), INGOs (international non-governmental organisations), GROs (grassroots organisations), CBOs (community-based organisations), and DONGOs (donor NGOs – created and owned by donors to do their job while shifting overhead costs outside), and MONGOs (my own NGO – an NGO that is the personal property of an individual often dominated by their own ego) (see Fowler 1997). Also important is a distinction made by the World Bank between operational NGOs (ONGOs) and advocacy NGOs (ANGOs). The ONGOs are primarily concerned with their own programmes on the ground and include international organisations (INGOs) headquartered in the North, as well as national organisations (NNGOs) operating in individual developing countries, and CBOs that serve a specific population group in a narrow geographic area. ANGOs are mainly concerned with advocating a specific point of view or concern and seek to influence the policies and practices of governments and other organisations, such as Greenpeace. For the World Bank, advocacy is viewed as a role for international rather than national NGOs (World Bank 1995). Overall, these variations stem from the diversity of NGOs as well as the growth in the number and types in recent years, not least in Middle America.

Another important type of civil society and NGO organising relates to the proliferation of non-state networks and actors that have emerged on the international scene in the last 20 years (Lipschutz 1992). Often forming part of global civil society, many have grown from local-level, bottom-up initiatives, achieving more formal status due to the increased visibility and recognition of their activities and the changing international climate. Many have emerged as a result of relations with inter-governmental organisations and especially with the agencies of the UN. Others have emerged and been developed around specific issues and grievances of a global or regional nature such as peace, human rights, development and ecological issues related to international fora such as the Rio Earth Summit, the Copenhagen Social Summit and the Beijing Women's Conference (Coate *et al.* 1996).

Why do NGOs exist?

As mentioned above, NGOs operate on a non-profit basis, deliver services, and often campaign for the rights of marginalised sectors of society. However, their existence warrants further examination in order to clarify what distinguishes NGOs from other social actors. In turn, this helps us to understand better why international donor organisations increasingly look to NGOs as key actors and why they channel funds through them. This also allows us to examine the limitations of NGOs.

In contrast to the usual moral explanations, Scott and Hopkins (1999) develop a useful economic model to explain why NGOs exist. In demonstrating why NGOs are superior to private firms and public agencies at supplying development goods and services, they suggest that NGOs operate in situations where there is an excess demand for goods and services usually provided by the public sector. This occurs in situations where the government of a country does not meet the basic needs of the population through lack of resources, lack of knowledge of needs, or incompetence, corruption or lack of political will in the sense of government failure. The reasons why NGOs provide these goods and services rather than private profit organisations relates, first, to notions of altruism – the concern for the well being of others – and second to better 'development technology' related to their superior knowledge of identifying people's needs and priorities. The comparative advantage of NGOs in terms of what makes them better than private organisations comes in part from employing altruists, since altruists are believed to work for lower wages or work harder for the same wages. However, these types of charitable acts are not merely a one-way process, but rather a price paid to receive a feeling of satisfaction, a lessening of guilt, or even public recognition (see Andreoni 1990). Scott and Hopkins (1999: 5) outline the advantages for altruists from working in NGOs; first, it provides a 'warm glow' – satisfaction derived from devoting effort in favour of the beneficiaries – and second a general altruism – the indirect increase in personal satisfaction resulting from an improvement in the beneficiaries' welfare. Therefore, as long as altruists continue to work within NGOs for these personal benefits, NGOs will continue to have a comparative advantage over private sector service deliverers. Overall, NGOs are seen to be better at understanding and representing the needs of the people with whom they work, given their closer working relationship.

What do NGOs do?

The traditional role of NGOs has been the delivery of services to alleviate the symptoms of poverty, although not necessarily its causes. This is usually in the fields of food, health, housing, education, production, credit and micro-finance and fostering self-reliance. However, over time, NGOs have developed multiple identities and activities. As a way of classifying these, Korten (1990) identifies four generations of development NGO strategy (Box 10.1). While these generations may be read as process or

---

Box 10.1 Generations of development NGO strategies

NGOs are involved in a wide range of activities that can be classified according to their orientation. Often, these activities evolve over time, from a starting point of the first generation.

First generation: involves relief and welfare, or the direct delivery of services to meet immediate deficiency. This is particularly relevant to emergency or humanitarian relief in times of crisis arising from 'natural' disasters or conflict.

Second generation: involves local self-reliance, or the development of the capacities of people to meet their own needs better.

Third generation: involves establishing sustainable development systems, or involvement in the policy formulation process of governments and multilateral organisations.

Fourth generation: involves political advocacy and campaigning in order to support people's movements and promote a broader social vision.

Source: Korten (1990).

progression, where involvement in the first generation activity of relief and welfare leads on to the second and so on, they may also be seen as an overlapping system. The roles of networking and research are also increasingly important in contemporary NGOs (Vakil 1997: 2063). In the past, NGOs have been criticised for failing to capitalise on their knowledge of grassroots realities in their dialogue with government and donor agencies (Clark 1992). With advocacy becoming an important strategic role of NGOs, networking and research are ever more crucial as legitimising this role through professionalisation in relation to the public, governments and official donor agencies.

How do NGOs operate?

The ways in which NGOs operate in relation to the state, private sector organisations and the international community are important in understanding a number of recent debates concerning NGO legitimacy. With private firms accountable to customers via market forces and governments accountable to their electorate, to what extent are private autonomous NGOs accountable to their users and beneficiaries? International donors act via NGOs if they believe they will act in the best interests of the intended beneficiaries. Beneficiaries work with NGOs since they believe that they will protect and promote their interests. If an NGO is not credible to either side then it will, in theory, cease to function. As NGOs are not necessarily 'democratic' in themselves, it raises the question of who represents what to whom in these relations. Attack (1999) suggests the need to consider four key issues related to NGO legitimacy. These include representation and accountability, values, participation and empowerment, and effectiveness.

Accountability is a critical issue for NGOs and concerns to whom NGOs are ultimately accountable and the contradictions and conflicts that may arise for national NGOs dependent on overseas donors. Edwards and Hulme (1995) discuss the question of 'multiple accountabilities', which on the one hand relates to partners or beneficiaries, staff and supporters (downwards), and on the other to various donors (upwards). The demands of donors on NGOs to prioritise certain activities or act in certain ways may weaken the very comparative advantage that causes these donors to invest in NGOs in the first place – their proximity to the users of development projects.

This issue of accountability is also interrelated with financing of NGOs. The funding needs of national NGOs inevitably means that they often depend on grants from international and Northern agencies. In a sample of NGOs in the Central America region, the organisations reported that among their principal sources of financing were the NGOs from the North (43 per cent of the sample), followed by public entities (39 per cent), contributions from affiliates (22 per cent), sale of services (18 per cent), international offical development agencies (17 per cent) (such as the Department for International Development), private businesses (14 per cent), multilateral organisations (15 per cent) (such as the United Nations Development Programme and Inter-American Development Bank), foundations (9 per cent), individuals (10 per cent) and others (12 per cent) (FACS 2000: 8). In the light of such reliance, national NGOs often have to compromise their primary aims in order to secure funding (Edwards and Hulme 1996). However, securing funding from international sources is becoming increasingly difficult in Middle America. As a result, many national and regional NGOs are having to consider alternative funding sources and activities, including self-financing through the diversification of services (Puntos de Encuentro 1998 on Nicaragua and Central America). Many national NGOs are being forced to introduce user charges or to sell their services, therefore increasing their reliance on market mechanisms. These changes may therefore undermine the ways in which NGOs function, raising questions about the fundamental characteristic of NGOs relating to their value systems.

The belief and value systems of NGOs are other key factors affecting their operation. Thomas (1992) suggests solidarity as the principle regulating NGOs in that they function on the basis of common interests rather than coercion or a desire to extract maximum profits. This notion of solidarity is also referred to as 'voluntarism' (Bratton 1989; Korten 1990), signifying a way of operating which is distinct from the other sectors (see also the Scott and Hopkins model mentioned earlier). These value systems help explain the growth in the importance of NGOs within contexts of government and market failure. However, it should also be noted that these values are being increasingly eroded, resulting in a severe identity crisis among NGOs as the principles and values of 'voluntarism' are replaced by the values of the market (Fowler 1997: 33). Others suggest the term 'voluntary' be excluded from definitions of NGOs as an acknowledgement of the increasing 'professionalisation' of the NGO sector and international recognition of their work (Vakil 1997: 2059).

In relation to participation, the ultimate benefit of development for many is the 'empowerment' of those disadvantaged groups in a society. While for some, empowerment is the key concept within the development discourse, for others it remains a vague and elusive term allowing for misuse and abuse (see Attack 1999; Edwards and Hulme 1992). One central aspect of the process of empowerment is participation. In recent years, many bilateral and multilateral agencies have embraced the idea of participatory development, which many NGOs have taken as central for a number of years (World Bank 1995; UN 1999). This increasingly places people at the centre of the planning process and has been high on the agenda of the international development NGOs working on poverty and environmental concerns. However, real participation as opposed to consultation, co-opting, or coercion is difficult to achieve and is a learning process in itself.

Effectiveness of NGOs and 'scaling up'

One of the greatest concerns in terms of NGO effectiveness is that, while they may be effective at a local level, at the international or macro-level their impact on policymaking is negligible. In general, they have largely failed to influence political regimes or to bring about fundamental changes in attitudes (Edwards and Hulme 1992). Instead, it is the bilateral and multilateral donors such as the World Bank and International Monetary Fund (IMF) which play much more dominant roles in determining the ideological policy regimes within which NGOs work. In the past, NGO lobbying has failed to change the structure of the world economy or the ideologies of its ruling institutions. For instance, multilateral organisations and NGOs still hold quite different views on the importance of economic growth in poverty reduction strategies. Indeed, official donors often criticise NGOs for doing little work on developing workable alternative policy proposals to the official ones they oppose. However, through their links with grassroots organisations and wider civil society, this may be changing through a combination of pressure from below and above, and through links with international NGOs (see below on Nicaragua).

Measuring the impact of an NGO is an inherently difficult task. Similarly, the effectiveness of their interventions is difficult to measure since they are rarely able to control the factors influencing the outcomes of their work (Edwards and Hulme 1995: 11). Perhaps for these reasons, emphasis on evaluation of effectiveness by donors has been somewhat lacking. Overall, the effectiveness of NGOs as agents of development has been exaggerated. Moreover, those viewed as effective usually operate only at a local level and have not been able to recreate this success or to 'scale up' (see Edwards and Hulme 1992).

However, this has been changing recently as more network and umbrella organisations are being formed. The shift in NGOs moving from working at the local level towards greater involvement in networks and policy advocacy depends not only on the political context in which the NGO operates and the internal characteristics of the NGO, but also on the influence of other NGOs and NGO networking (Fisher 1998). In relation to this, there are also practical and theoretical limitations to group networks and sizes and their abilities to perform (Olson 1982). In addition, there are two major requirements for effective lobbying: first, a degree of openness on the part of the organisation being lobbied, and second, a thoughtful strategy on the part of the network to target efforts. The heterogeneity of NGOs also acts as a potentially limiting factor. Indeed, relations between NGOs within networks are often described as encompassing 'co-operation and conflict'. While NGOs and civil society may organise and co-operate in order to campaign towards a common goal, conflict over how best to achieve this goal – and even agreeing what the central goal should be – will always present challenges. If interventions by international organisations to 'strengthen' civil society via NGOs do not recognise these issues then only superficial co-operation can occur.

Despite these problems, regional NGOs and regional NGO networks are growing in importance in Middle America, especially in relation to the integration and reconstruction process post-'Mitch'. The following sections will examine more closely the more 'political' advocacy role of NGOs, as well as the links between NGOs, civil society and governments.

NGOs, civil society and network relations in post-'Mitch' Middle America

The participation of civil society actors, who have been the representatives of the social groups most seriously affected by policy decisions, has historically been largely absent from decision-making processes. This has created what some authors have called 'a significant democratic deficit' (Serbin 1998). Many differing endogenous national civil society initiatives and expressions have emerged in response to this. NGOs are playing an active role in the development of civil society in Middle America at the national and subregional levels through their horizontal and vertical networking abilities. In particular, these NGOs link relations between grassroots organisations and the international community. Many national NGOs form an integral part of an organised national civil society and are trying to build stronger national, regional and global alliances in an attempt to influence international global political agendas and decision-making. One of the main strategic roles that NGOs can play in political and economic development is through their ability to energise and activate networks to push for social transformation through the formation of spatial co-ordinations at national, regional and international level.

Network structures of civil society and NGOs at national and regional levels

The organisational structures of national and regional civil society and NGO networks vary, depending on the country, in relation to their historical, cultural, political and economic development, and with regard to internal and external power relations (Hengstenberg *et al.* 1999). National co-ordinating organisations in the countries of the region tend to be organised into sectoral networks around specific themes and activities such as health, education, democracy, gender and small business development. The national networks are not themselves NGOs in the strict organisational sense, but rather co-ordinating bodies of a variety of types of organisations. Some of these include NGOs, as well as social movements, sectoral networks, territorial networks, producers' associations, unions and federations. These national sectoral networks are often combined in different ways into national intersectoral co-ordinating bodies.

In Central America, regional-level networks tend to be organised around sectoral interests, such as the *Desarrollo Sostenible de los Asentamientos Humanos en Centroamérica* (Sustainable Development of Human Settlements in Central America – CERCA) focusing on settlement issues and comprising 27 member organisations from a number of countries. Other initiatives with a broader focus include the *Iniciativa Civil para la Integración Centroamericana* (Civil Initiative for Central American Integration – ICIC), which is composed of 12 organisations concerned with the integration process in the region. In the Caribbean, the Caribbean Policy Development Centre (CPDC) groups more than 21 organisations of a regional, subregional and national nature. Together with labour unions and the business sector, the CPDC was incorporated as the third 'social partner' by CARICOM heads of government in their regional Consultative Council. This may be taken as a sign of the greater ability of such co-ordinating bodies to gain an effective voice for civil society in decision-making processes than single local initiatives.

Networking of organised civil society in a number of Middle American countries is strong at a national level and is beginning to improve at a subregional level, especially around trade and integration in the Caribbean and as a response to hurricane 'Mitch' in Central America. However, initiatives to link the countries of the wider Middle America region remain weak. Again, sectoral networking appears to be most established as the national sectoral networks also combine into regional sectoral networks, such as the *Comité Regional de Promoción de Salud Comunitaria* (Regional Committee for the Promotion of Community Health – CRPSC), which covers Mexico, Central America and the Caribbean (CRPSC 1999). Other initiatives at the Middle America level include attempts at organising civil society to influence the more recent regional state-level grouping of the Association of Caribbean States (ACS), which includes governments of the greater Caribbean region. As part of this initiative, 1997 saw the first 'Forum of Civil Society in the Greater Caribbean', which included representatives from Central America, the Caribbean and Venezuela. However, the fact that only 41 participants attended suggests either funding constraints, lack of interest in initiatives at this 'super'-regional level, or lack of association with this concept of 'region' (for a list of participants see CRIES-INVESP 1998).[2] The following discussion will focus on the subregional initiatives of organised civil society as most advances are being made at this level.

Regional level civil society initiatives post-'Mitch'

In the post-'Mitch' context, a number of national and regional co-ordinations bringing together diverse sectors of civil society have emerged, with the aim of achieving greater participation in the reconstruction process. This process began in November 1998, immediately after hurricane 'Mitch', when regional civil society activated and formulated a declaration to lobby for a process of reconstruction based on sustainable human development with the maximum participation of civil society. Although initially excluded from the first meeting of the Consultative Group[3] on Central America in Washington, D.C. in 1998 between international donors and Central American governments, representatives of organised civil society undertook a successful lobbying campaign to be allowed to participate in such meetings in the future (see Box 10.2 for a chronology of meetings and events).

Since Washington, D.C., national-level co-ordinations of civil society have continued to meet regionally to develop proposals to be included in national and regional reconstruction plans, as well as developing internal structures and lobbying mechanisms to permit an active participation of civil society in the reconstruction process. These national-level co-ordinating organisations include the following: the *Coordinadora Civil para la Emergencia y Reconstrucción* (Civil Co-ordinator for Emergency and

[2] Although the concept of Middle America is useful for academics and theorists, it is not used by the people of these countries. Local initiatives to organise at a regional geographical level usually work within the concept of the 'Greater Caribbean' rather than 'Middle America'.
[3] The Consultative Group meetings are co-ordinating fora between countries that receive aid and the international donor community. They permit co-operating countries to influence policies of countries receiving international support and implementing SAPs.

Plate 10.1   One week after hurricane 'Mitch' in a refugee camp in Tuskru Taram
Río Coco, Nicaragua, November 1998
Photo: Liz Light

Reconstruction – CCER) in Nicaragua; *Espacio INTERFOROS* (Space
Interforos) in Honduras; *La Instancia de Seguimiento al Grupo Consultivo*
(Consultative Follow-up Group) in Guatemala; *El Foro de la Sociedad
Civil por la Reconstrucción y el Desarrollo* (Forum of Civil Society for Recon-
struction and Development) in El Salvador; *Centroamerica Solidaria*
(Central American Solidarity) in Costa Rica, together with the regional-
level organisations, *Iniciativa Civil para la Integración Centroamericano* (Civil
Initiative for Central American Integration – ICIC) and the *Coordinadora
Centroamericana del Campo* (Central American Rural Co-ordination). All
these organisations have elaborated policy proposals for the reconstruction and transformation of the region (for summaries of these proposals
see ALFORJA 1999).

In a series of meetings between January and April 1999, these co-
ordinating institutions met and negotiated some key issues to be included
in the second Consultative Group meeting in Stockholm in May 1999,
between the regional governments and the international co-operating governments of the official donors where funding priorities were decided (see
Box 10.2). As a result, not only was organised civil society's voice heard at
the Stockholm meeting, but their proposals for reconstruction were incorporated into the official declaration of the Consultative Group meeting
in Stockholm. The agreements of the meeting were encompassed in the
Declaration of Stockholm, which recognised that reconstruction must take
place through a co-ordinated effort based on the priorities of each country.
These priorities were: reducing environmental and social vulnerability;
consolidating democracy and good governance through decentralisation

---

Box 10.2   Consultations and key CCER events in Central America, post-'Mitch'

Date	Event
October 1998	Hurricane 'Mitch' hits Central America
December 1998	Emergency Consultative Group meeting for Central America in Washington, D.C.
February 1999	Phase I of the Social Audit undertaken in Nicaragua by the CCER
April 1999	First National Meeting of Civil Society in Nicaragua hosted by the CCER
April 1999	First Regional Meeting of Civil Society held in Honduras hosted by INTERFOROS in Tegucigalpa
May 1999	Consultative Group Meeting in Stockholm, Sweden
September 1999	Phase II of the Social Audit undertaken in Nicaragua by CCER
November 1999	Anniversary Meeting of Civil Society in Nicaragua hosted by the CCER
November 1999	Second Regional Meeting of Civil Society held in Nicaragua hosted by the CCER in Managua
February 2000	Government of Nicaragua cancels Consultative Group meeting for Nicaragua planned for Managua
February 2000	National Civil Society Forum in Nicaragua hosted by CCER, evaluates completion of Stockholm agreements
May 2000	Consultative Group Meeting held in Washington, D.C. evaluates completion of Stockholm agreements

---

of power and citizen participation; promoting and respecting human rights with the rights of children, ethnic and other minority groups and the promotion of gender equality being specifically highlighted. These were to take place within a context of transparency and accountability of all actors, accompanied by a reduction in the debt burden. Funding from international official donors and the international NGO solidarity donors was agreed, although more progress has been made with the latter. This has been due to problems relating to transparency and good governance by the regional governments, especially in terms of corruption.

As well as two interim meetings, representatives of national and regional civil society networks that participated in Stockholm met one year after 'Mitch', in Managua, Nicaragua, at the Second Regional Meeting of Central American Civil Society hosted by the CCER. The intention was to establish a space to evaluate Stockholm and develop follow-up agreements. This time other key themes were discussed, such as regional integration, globalisation and democratisation, along with the role of popular social movements (Segundo Encuentro Regional 2000).

Within each country, civil society also developed follow-up initiatives. Guatemala and Costa Rica developed national co-ordinating processes as well as advocacy activities. In El Salvador, formal mechanisms of consultation between civil society and the government were established, facilitated by the United Nations Development Programme (UNDP). In Honduras, *Espacio INTERFOROS* has been in constant discussion with the government over international co-operation on the need for transformative reconstruction, despite lack of progress in opening real spaces for civil participation in the definition and management of new policies, programmes and projects (INTERFOROS 1999). It has also managed to secure some flexibility in conditions attached to the Highly Indebted Poor Countries (HIPC) initiative (see McIlwaine, this volume). Overall, these developments highlight the important role of these co-ordinating organisations in influencing the shape of the reconstruction process in Central America post-'Mitch'.

A case study of the Civil Co-ordinator for Emergency and Reconstruction (CCER) in Nicaragua

In Nicaragua, the last 20 years have witnessed the development and strengthening of different expressions of organised civil society, with the rise of organisations working within the themes of health, education, the environment, human rights, governability and gender. Hurricane 'Mitch' brought with it a new stage in the development of these distinct organisations as they came together with the objective of co-ordinating a collective effort to respond to the immediate necessities of the populations in the affected zones (CCER 1999a). Out of this, the CCER was formed as a co-ordinating body of 21 networks, representing the involvement of more than 350 national NGOs, social movements, sectoral networks, producer associations, unions, collectives and federations.[4]

The CCER was not created or imposed from above by agencies of the international community, as was the case in other Central American countries post-'Mitch'. Instead, it emerged from below, out of existing national and local networks. Although it was initially responding to the immediate needs of those affected by 'Mitch' in terms of providing food, water, clothing, housing and medical help, this role evolved into a more strategic one. The collective experience and knowledge of the CCER participant organisations, together with a lack of confidence in the ability or willingness

---

[4.] The CCER is composed of the following national networks: *Asociación de Mujeres 'Luisa Amanda Espinoza' – AMNLAE; Comité Costeño de Apoyo a la gestión de Emergencia y Rehabilitación en la Costa Caribe Nicaragüense; Consejo de la Juventud de Nicaragua – CJN; Coordinadora Nicaragüense de ONGs; Coordinadora Nicaragüense de ONGs que trabajan con la Niñez y la Adolescencia – CODENI; Federación de Coordinadora Nicaragüense de Organismos por la Rehabilitación e Integración – FECONORI; Federación Organizaciones No Gubernamentales de Nicaragua – FONG; Foro de Educación y Desarrollo Humano – FEDH; Grupo de Coordinación para la Prevención del Consumo de Drogas; Grupo FUNDEMOS; Grupo Propositivo de Cabildeo e Incidencia – GPC; MIPYMEs; Movimiento Comunal Nicaragüense – MCN; Movimiento Pedagógico Nicaragüense; Red de Mujeres Contra la Violencia; Red de Mujeres por la Salud 'María Cavalleri;' Red Nicaragüense de Comercio Comunitario; Red Nicaragüense por la Democracia y el Desarrollo Local; Red de Vivienda; Unión Nacional de Agricultores y Ganaderos – UNAG; Unión Nicaragüense de Campesinos Agropecuarios – UNCA.*

257

of the government to undertake a real process of reconstruction, resulted in the recognition that civil society had to propose its own plan for reconstruction, and not merely comment on the plans of others. Thus, while the immediate response of organised civil society to 'Mitch' was to adopt a 'welfare' role, it used its national co-ordinating networks to develop policy proposals and quickly progressed to a 'political advocacy' role in relation to the government and international donor community.

Government–civil society relations in Nicaragua: co-operation and conflict

By the time of the emergency meeting of the Consultative Group for the Reconstruction of Central America in Washington, D.C. in December 1998, the CCER had outlined its own proposal for the reconstruction of the country, based on a recognition of the need for organised civil society to move beyond campaigning 'against', and rather to campaign 'for' (CCER 1998). The CCER quickly developed its role from one of co-ordinating the delivery of services to alleviate the situation, to one of discussing the real causes of the situation.

The proposal, written by representatives from the different organisations that form the CCER, stressed the need not only to rebuild the damaged infrastructure of the country and reply to the basic needs of those affected, but to improve the conditions of the most vulnerable in society with a shared vision of sustainable human development. This proposal represented the first achievement of the CCER in a number of ways. First, that a shared proposal was produced at all in such a short space of time was an achievement, given the diversity of participant organisations in the CCER. Second, as the only expression of civil society to arrive at Washington, D.C. with a formulated proposal document, the CCER won recognition from donor agencies and other governments as a legitimate actor in the reconstruction process in Nicaragua. This legitimacy was then recognised by the national government post-Washington, D.C. with the formation of the National Council for Economic and Social Planning. This allowed for the official participation of representatives from civil society in the working groups created by the government to write the national reconstruction plan. It is important to note, however, that while officially formed in February 1999, it did not begin to function until November. Furthermore, the official government plan for reconstruction was not discussed with civil society via this officially created space, nor more generally, not least because it was written in English and then translated into Spanish (the national language) just before the second international meeting to discuss the reconstruction of the region. The government's focus on reinstating the damaged infrastructure of the country, particularly a road-building programme, was attacked by the CCER as being unable alone to bring about the social transformation desired (CCER 1999d).

Post-Washington, D.C., the CCER had recognised the need to validate its own proposal, thus far produced by a small group of representatives from different participant organisations, and to improve it. The CCER used two methods to produce a more inclusive document. First, themed commissions were formed to rework and develop the proposal in the key areas. These consisted of seven commissions on health, education, production and small business, environment and development, decentralisation and local power, housing and infrastructure, and the macro-economic and debt commission. These were complemented by the Gender Commission, and commissions focused on young people and children. In addition, due to the differing ethnic identities of the autonomous Atlantic coast regions, there was a semi-autonomous commission of the Caribbean Coast.[5] These transversal commissions worked with the themed commissions to ensure that issues such as power relations between men and women, and ethnicity, were taken into account in the proposals. Once a draft document had been written, a consultation process began (CCER 1999b). The document was discussed via the networks at 18 open meetings covering the different regions of the country, plus four themed meetings at the national level, representing 376 organisations. This process culminated in the First National Meeting of Civil Society with the participation of over 1,500 delegates from civil society organisations.

At the Consultative Group meeting in Stockholm in May 1999, the CCER presented a proposal for the reconstruction and transformation of Nicaragua that was based on, if still not a truly participatory, at least a truly consultative process. It sought not only to challenge the government document but to counteract this focus. It highlighted the need to address the underlying causes of vulnerability, such as unequal power relations and resource distribution, in order to mitigate the impact of events such as 'Mitch' in the future. The role of civil society in the Stockholm meeting had also changed since the Washington, D.C. meeting. The CCER and its sister organisations in the other countries of the region were given, for the first time at such meetings, official recognition and participatory status.

Stockholm saw common agreements between expressions of civil society from the North and the South that successfully entered the official declarations of the meeting, accepted by the governments of the region. The meeting in Stockholm was also important since it looked set to mark a new era of dialogue between civil society and national governments. Some achievements were made, most notably acceptance by all sides of a proposal for indicators to evaluate the reconstruction process produced by a team of consultants. However, the agreements for more discussion and participation in Nicaragua did not last.

An analysis of the activities of the CCER after Stockholm may help to explain the rapid deterioration of the relations between the government and the CCER, highlighting the fragility of state–civil society relations. First, while CCER's advocacy role was always central, this did not negate others. The CCER had recognised the need to campaign from a basis of evidence, and thus a research role had been adopted almost from its initiation. While a number of research projects exist within the CCER, the Social

---

[5.] The Caribbean coast of Nicaragua, as in Central America as a whole, is characterised by high proportions of Afro-Caribbean and indigenous Indian populations, with the former migrating at the turn of the century from various Caribbean islands to work on the banana plantations. Historically, the region has thus been isolated and culturally distinct from the rest of the country.

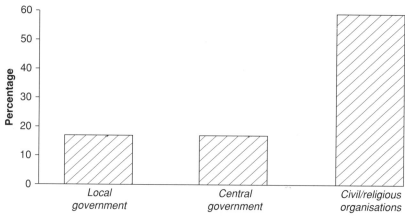

Figure 10.1 Perception of what the central government is doing in response to the damage caused by hurricane 'Mitch' (percentage of interviewees) *Source*: CCER (1999e).

Audit project is one of the most important. This allows civil society to generate its own information on the situation in the country and monitor and evaluate its own and the government's efforts in reconstruction activities. To date, two phases of this auditing process have been completed, each consisting of interviews with over 6,000 men and women (including both young and older people), covering the municipalities worst affected by 'Mitch'. While the first audit addressed the emergency phase (CCER 1999c), and the second, the reconstruction phase (CCER 1999e), both had a shared focus.

Information has been gathered not only on practical issues, such as the damage suffered, needs and priorities of those affected, amount of aid received and from whom, but also on the distribution of aid, issues of fairness, transparency and accountability and levels of participation in planning and decision-making processes. The Social Audit shows that national and international NGOs were the most active in reconstruction and that their projects were seen to be most beneficial to the people affected (59 per cent) compared with local and national government sectors (17 per cent respectively). When asked to identify the most important thing the government was doing in the reconstruction process, 60 per cent of those interviewed replied 'nothing' (CCER 1999e). While these results suggest that all sectors, including the NGOs, could improve, they are particularly damning in terms of the population's views of the government (see Figure 10.1).

With the shift from the emergency to reconstruction stages, funding varied between the government and NGOs. In the early stages of the emergency, more aid was channelled through civil society organisations (US$16 million) than through the government sector (US$12 million until November 1998) (Linneker *et al.* 1998). Planned external reconstruction funds (composed of donations and loans) being channelled through the government sector between 1999 and 2003 should total US$1.2 billion.

260

Table 10.1 Reconstruction funds channelled into Nicaragua in 1999 (percentage)

Activity	Government	NGOs	*Other NGO funds	Total NGOs	Total
Production and finance	34	28	n.a.	n.a.	n.a.
Infrastructure	36	13	n.a.	n.a.	n.a.
Governance	6	6	n.a.	n.a.	n.a.
Social sector	24	53	n.a.	n.a.	n.a.
Total US$ (millions)	US$614.1	US$154.4	US$145.6	US$300	US$914.1
Percentage	67.2			32.8	100

Sources: Authors' estimate from official and unofficial sources (Gobierno de Nicaragua 1999, 2000a; CCER communication).
Notes
* Direct bilateral and multilateral organisation funds channelled through NGO sector
n.a. = not available

There is a planned gradual fall from US$614.1 million in 1999 to US$38.8 million in 2003 (Gobierno de Nicaragua 1999). While information on future funding commitments through the NGO sector are less certain, estimates for external reconstruction funds channelled into Nicaragua in 1999 through the government and NGO sectors show that the NGO share varies between 20 and 33 per cent, depending on estimates used. In total, approximately US$914.1 million of external funding was channelled through both the national government and NGO sectors in 1999 (see Table 10.1).

In the reconstruction stage, a larger proportion of NGO spending is being undertaken in the social sector (such as health, housing and education) while government expenditure tends to be on physical infrastructure, and production and finance. These funds should be seen in relation to an estimated GDP of US$2,190 million in 1999 (FIDEG 2000) and represent 41 per cent of this total. These estimates ignore the changing distribution between individual donor countries, since some countries channel the majority of their external funding through the NGO sector due to their concerns about transparency and corruption problems in relation to the government sector. Nor do they account for recent withholding of funding by some donor countries to government agencies for similar reasons. Institutionalised corruption is a major concern in Nicaragua, with 85 per cent of the general public perceiving government ministers, political leaders and members of the National Assembly to be corrupt (CCER 2000a: 40).

The channelling of funds through NGOs and the questioning of the government role in reconstruction through the presentation of the Social Audit results at the Anniversary National Meeting of Civil Society, hosted by the CCER in November 1999, may not have helped CCER relations with the government. Further problems arose as organised civil society, including and at times led by the CCER, protested at what they saw as threats to democracy and good governance. The key issues include the pact between the ruling government party, the *Partido Liberal Constitutionalista* (Liberal Constitutional Party – PLC) and the opposition – the *Frente Sandinista Liberación Nacional* (Sandinista National Liberation Front – FSLN), as well as constitutional reforms, the arrest and imprisonment of the head of the independent audit commission, and the expansion of government control over this body. Actions by civil society took the form of street demonstrations and lobbying outside government offices, through to public announcements and press conferences. A campaign to petition for a referendum on the reforms to the constitution was also launched.

Deterioration in relations continued as the CCER criticised the government's strategy document for poverty reduction (a necessary component of being accepted on to the heavily indebted poor countries list – the HIPC initiative – see McIlwaine, this volume). This called into question the evidence presented on the situation in the country and the capacity for the policies presented to surmount the enormity of the problems faced. The government had also produced an evaluation document (Gobierno de Nicaragua 2000b), highlighting its achievements since Stockholm, to be presented at the meeting planned for February 2000. The cancellation by the government of this meeting due to a prior engagement of the President to attend a meeting in Mexico caused a further rift. This was made worse by the decision of the CCER to hold in its place a meeting of civil society to evaluate the reconstruction of the country since Stockholm. This meeting was attended by over 300 people, representing 200 expressions of civil society. This can be read as marking a key stage in the development of the CCER as a legitimate participatory space and representative voice of civil society.

A progression or development is apparent in terms of civil society proposals from emergency to reconstruction. These began with a document produced by 'experts' in Washington, D.C. in 1998, followed by a document based on the work of themed commissions validated through a wide consultation process in Stockholm, moving to a document produced using participatory methods to assess the situation in the country. In turn, this was complemented by a more direct critique of the government document produced by the themed commissions in the run-up to the rescheduled follow-up to the Stockholm meeting in Washington, D.C. in May 2000. Thus the CCER could be said to be working actively to improve its legitimacy and credibility with both donor governments and agencies, and the people it seeks to represent.

The February 2000 meeting was also important as marking a clear change in the relations between organised civil society as represented by the CCER and the government to one of open confrontation. This confrontation initially took the form of public attacks, via the national media, not on the legitimacy of the CCER in itself but rather on its official spokesperson, Ana Quirós Víquez. While the government campaign began in this way, threatening to revoke the status of Ana Quirós Víquez as a nationalised Nicaraguan (i.e. a person born outside the country in Mexico but who has chosen and been accepted to take the nationality of that country), it quickly took a more political focus.

This situation may be considered within the 'paradox of civil society' outlined above, where a strong politically independent civil society might put forward excessive demands on a weak government that are not consistent with democracy and governability. The Nicaraguan government stance has two key strands. First, it challenged the right of the CCER to involve itself in 'political' issues. This refers to both the CCER's criticism of government documents and strategies (especially the poverty reduction strategy) and its involvement in the campaign for a referendum on constitutional reforms. Comments by some members of the government (and some representatives of the opposition) suggest that they view political parties as the true representatives of civil society in the 'political' realm of policy formulation. On the referendum campaign, the government focus on the campaign as 'political' in the 'party political' sense, the CCER states that its mission includes campaigning on issues of transparency, accountability and good governance (agreed in the Declaration of Stockholm), arguing the lack of consultation on the reforms to be a legitimate focus for its work (CCER 2000b).

The basis of the confrontation could be perceived as a difference in opinion over the role of civil society organisations. However, an alternative reading may suggest the confrontation to be a response to the perceived lack of 'success' of the government in terms of meeting the agreements of Stockholm (as evidenced by the Social Audit results), and the 'success' of the CCER in its role in highlighting this. Moreover, the CCER may be seen to have been 'successful' in presenting alternatives due to its proactive role, for example, going beyond merely criticising government strategy and presenting alternative plans.

Furthermore, the government moved beyond questioning the legitimacy of the CCER to questioning the legitimacy of the NGO sector in general. Recently, the government has presented plans to audit all NGOs and is proposing to carry out an audit on service delivery to be used as a comparison to the Social Audit. That these proposals have been met with some concern by the NGOs operating in the country is perhaps not surprising given the relations between the government and civil society, and the present government's repressive stance. However, all the activities proposed by the government in themselves should be welcomed by a civil society focused on issues of transparency and accountability.

Conclusions and future developments

While national NGOs have demonstrated a high degree of success in their traditional roles of local-level emergency and poverty alleviation, in relation to weak governments, there remains scope for improvement. Their wealth of local knowledge is often under-used in terms of improving responsiveness to needs and effective service provision. However, their real effectiveness in newer and increasingly important roles, especially advocacy or lobbying at national and international levels, is not clear-cut. An analysis of organised civil society in post-'Mitch' Central America has helped to highlight some of the achievements and shortcomings in this area.

The extent to which hurricane 'Mitch' acted as a catalyst for the energis-
ing of previously existing networks of both national and local organisations
is apparent. Co-ordinating bodies of civil society organisations throughout
Central America rapidly came together around a common desire to participate actively in the reconstruction and transformation of their countries
and the region. These diverse organisations were united by the perceived
need for an alternative (non-government) mechanism to articulate the views
of their participants to the national governments and international community. NGOs and civil society of the region may be judged as highly effective
in opening up spaces and creating mechanisms for discussion, especially
on the international level. The issue of their real ability to represent and
articulate the views of the grass-roots is less clear, since it depends largely
on their own internal mechanisms of consultation. The case study of the
CCER might suggest that this issue has been taken on board as a serious
concern by these co-ordinating bodies, important for improving legitimacy
both in the eyes of the people they seek to represent and international
agencies.

One contradiction to emerge in this process is between civil society's
ability to get policy demands heard via lobbying at an international level,
and agreements formalised via international pressure, and the lack of
ability to ensure that national governments carry them out. While it is
debatable whether lack of action by national governments stems from
inability, due to finance for example, or unwillingness, it highlights that
civil society co-ordinating bodies are reliant on governments. That organised civil society cannot replace all the functions of the government stems
first from more practical considerations and the lack of evidence on the
part of NGOs to 'scale up' their approaches to service delivery successfully. Second, it highlights issues of the internal contradictions that exist in
terms of the real representability, accountability and legitimacy of these
non-governmental groupings.

At times the coming together of value and fact can be a very explosive
social mixture; this has often been the experience in social relations in the
countries of the region. However, reciprocal learning relationships between
government and civil society are hopefully a valid and achievable goal,
but one that is dependent on the will of both sides to be open to debate
and discussion, and to address their own internal strengths and limitations.

For some authors, recent changes in social and spatial network relations at national, regional and international levels represent progress in
the development of a global civil society (Lipschutz 1992; Serbin 1998). The
recent disruption of the World Trade Organization (WTO) conference in
Seattle may stand as an example for the future. The Seattle demonstration
of global organised civil society owed its success to a combination of local
and global actions, through the ability to easily bring together people from
many different countries both physically in the city and 'virtually' on the
WTO website via modern communication technologies. While global alliances can be quickly built around certain issues, aided by the ease of
generating and disseminating information, this also highlights the technical
vulnerabilities of those international organisations and networks that also
depend on them. However, civil society's real effectiveness in changing

global power relations, policy regimes and dominant ideologies remains to be seen.

Useful websites on the Nicaraguan and Central American situation

www.ccer-nic.org: La Coordinadora Civil para la Emergencia y la Reconstrucción (Civil Co-ordinator for Emergency and Reconstruction – CCER), Nicaragua.

www.puntos.org.ni: Fundación Puntos de Encuentro, Nicaragua (a feminist organisation).

www.sdnnic.org.ni: La Red de Desarrollo Sostenible de Nicaragua (The Sustainable Development Network of Nicaragua – SDNNIC).

www.facs.org.ni: Fundación Augusto Cesar Sandino, Nicaragua.

www.nicanet.com.ni/fundemos: FUNDEMOS, Nicaragua.

www.cisas.org.ni: Centro de Información y Servicios de Asesoría en Salud (Centre for Information, Services and Assistance in Health – CISAS), Nicaragua.

www.cepredenac.org: El Centro de Coordinación para la Prevención de los Desastres Naturales en América Central (the Centre for the Prevention of Natural Disasters in Central America – CEPREDENAC).

www.osso.univalle.edu.co/tmp/lared/lared.htm: La Red de Estudios Sociales en Prevención de Desastres en América Latina (the Network of Social Studies of the Prevention of Disasters in Latin America).

www.arias.or.cr/Eindice.htm: Fundación Arias.

www.disaster.info.desastres.net/crid/index.htm: Disaster Info.

www.solidaridad.org: Costa Rica Solidaridad (Solidarity).

Further reading

Edwards, M. and **Hulme, D.** (eds) (1992) *Making a Difference: NGOs and Development in a Changing World*. Earthscan, London.
Comprises a series of short, useful articles on the roles and functions of NGOs. Includes some case studies from Middle America.

Fowler, A. (1997) *Striking a Balance: a Guide to Enhancing the Effectiveness of Non-Governmental Organisations in International Development*. Earthscan, London.
A key text on the function and effectiveness of NGOs.

Hulme, D. and **Edwards, M.** (eds) (1997) *NGOs, States and Donors: Too Close for Comfort?* Macmillan in association with Save the Children, Basingstoke.

An up-to-date focus on the challenges facing NGOs and their relationships with state and international donor organisations. Again, it includes case studies from Middle America.

Korten, D. (1990) *Getting to the 21ˢᵗ Century: Voluntary Action and the Global Agenda.* Kumarian Press, West Hartford, CT.
Provides useful definitional discussions on NGOs.

MacDonald, L. (1997) *Supporting Civil Society: the Political Role of Non-Governmental Organisations in Central America.* Macmillan, Basingstoke.
Outlines the theoretical positions on civil society, as well as case studies from Costa Rica and Nicaragua.

McIlwaine, C. (1998) Contesting civil society: reflections from El Salvador, *Third World Quarterly* **19**(4): 651–672.
A case study of El Salvador that also includes a discussion of the theoretical approaches to civil society. Adopts an explicitly geographical perspective.

References

ALFORJA (1999) Propuestas de la sociedad civil Centroamericana para la reconstrucción y transformación de América Central luego del Huracan Mitch, alforja Programa Regional Coordinado de Educación Popular, Costa Rica, May 1999.

Andreoni, J. (1990) Impure altruism and donations to public goods: a theory of warm-glow giving, *Economic Journal* **100**(401): 464–477.

Attack, I. (1999) Four criteria of development NGO legitimacy, *World Development* **27**(5): 855–864.

Bratton, M. (1989) The politics of government–NGO relations in Africa, *World Development* **17**(4): 569–587.

CCER (1998) Turning the Mitch tragedy into an opportunity for the human and sustainable development of Nicaragua. Washington, D.C. document, Coordinadora Civil para la Emergencia y la Reconstrucción (CCER), Managua.

—— (1999a) *Proposal for the reconstruction and transformation of Nicaragua: converting the tragedy of Mitch into an opportunity for the sustainable human development of Nicaragua.* CCER, Carqui Press, Managua.

—— (1999b) *Executive summary: proposal for the reconstruction and transformation of Nicaragua: converting the tragedy of Mitch into an opportunity for the sustainable human development of Nicaragua.* CCER, Carqui Press, Managua.

—— (1999c) *Social audit for the emergency and reconstruction: Phase 1 – April 1999.* CCER and CIET International, Carqui Press, Managua.

—— (1999d) Brief comparison of the proposals of the CCER and government of Nicaragua: notes for conference delegates. CCER, Managua.

—— (1999e) Auditoría social para la emergencia y la reconstrucción – fase 2 – noviembre. CCER, Managua.

—— (2000a) Balance general de la CCER sobre el cumplimiento de los objectivos, prioridades y principios definidos en la declaración de Estocolmo. CCER, May 2000, Managua.

—— (2000b) Press release, 17 February 2000, CCER, Managua.

CEPAL (1999) Evaluación de los daños ocasionados por el huracán Mitch, 1998; Sus implicaciones para el desarrollo económico, social y el medio ambiente, Naciones Unidas, Centroamérica, May 1999, Mexico.

CINDI (1999) Information from the website of the Centre for the Integration of Natural Disaster Information (CINDI) (*www.cindi.usgs.gov/*)

Clark, J. (1992) Policy influence, lobbying and advocacy, in Edwards, M. and Hulme, D. (eds) *Making a Difference: NGOs and Development in a Changing World*. Earthscan, London: 187–221.

Coate, R., Alger, C., and **Lipschutz, R.** (1996) The United Nations and civil society: creative partnership for sustainable development, *Alternatives* **1**(1): 93–122.

CRIES-INVESP (1998) *1er. foro de la sociedad civil del Gran Caribe, Coordinadora Regional de Investigaciones Economicas y Sociales (CRIES) y Instituto Venezolano de Estudios Sociales y Politicos (INVESP)*. Epsilon Libros, Caracas.

CRPSC (1999) *Reformas de salud en Centroamerica; vistas desde la perspectiva de la salud comunitaria, Comité Regional de Promoción de Salud Comunitaria (CRPSC)*. Carqui Press, Managua.

Edwards, M. and Hulme, D. (eds) (1992) *Making a Difference: NGOs and Development in a Changing World*, Earthscan, London.

—— (1995) *Non-Governmental Organisations – Performance and Accountability: Beyond the Magic Bullet*. Earthscan, London.

—— (1996) Too close for comfort? The impact of official aid on nongovernmental organisations, *World Development* **24**(6): 961–973.

FACS (2000) Cambios institucionales ONG: los retos y tareas de las ONGs a partir del año 2000, XX Aniversario, Encuentro Internacional, Base Documento, Fundación Augusto Cesar Sandino (FACS), Managua.

FIDEG (2000) *El observador económico*, No. 98, marzo-abril, 2000.

Fisher, J. (1998) Promoting democratization and sustainable development, in Fisher, J. (ed.) *Nongovernments: NGOs and the Political Development of the Developing World*. Kumarian Press, West Hartford, CT: 105–134.

Foley, M.W. and **Edwards, R.** (1996) The paradox of civil society, *Journal of Democracy* **7**(3): 35–52.

Fowler, A. (1997) *Striking a Balance: a Guide to Enhancing the Effectiveness of Non-Governmental Organisations in International Development*. Earthscan, London.

Fundación Arias (1997) Diagnóstico sobre la incidencia en Centroamérica: proyecto la formación de una cultura democrática en Centroamérica, el papel socializador de las ONG, Fundación Arias para la Paz y el Progreso Humano, San José.

—— (1998) Marco jurídico que regula a las organizaciones sin fines de lucro en Centroamérica, serie: el derecho y la sociedad civil No. 3. Fundación Arias para la Paz y el Progreso Humano, San José.

Gobierno de Nicaragua (1999) Recursos contratados y en gestión, disponibilidades de recursos externos por sector y sub-sector 1999–2003 donaciones y prestamos, Presidencia de la República, Secretaria de Cooperación Externa, 20 July 1999.

—— (2000a) Montos de cooperación recibida via ONGs, Secretaria de Cooperación Externa, January 2000.

—— (2000b) El desarrollo del diálogo y diálogo del desarrollo: sociedad, gobierno, ayuda externa en Nicaragua en el nuevo milenio, Version 19 January 2000.

Hengstenberg, P., Kohut, K. and **Maihold, G.** (1999) *Sociedad Civil en América Latina: Representación de Intereses y Gobernabilidad.* Asociación Alemana de Investigación sobre América Latina (ADLAF), Friedrich Ebert Stiftung-FES, Editorial Nueva Sociedad, Caracas.

INTERFOROS (1999) Propuesta de reconstrucción y transformacion de Centroamérica: declaracion de las coordinadoras nacional y las redes regionales, producto del Encuentro Regional de la Sociedad Civil por la reconstrucción y el desarrollo, Tegucigalpa, Honduras, 21–22 April.

Karl, T.L. (1995) The hybrid regimes of Central America, *Journal of Democracy* **6**(3): 72–86.

Korten, D. (1990) *Getting to the 21st Century: Voluntary Action and the Global Agenda.* Kumarian Press, West Hartford, CT.

Linneker, B.J., Quintanilla, M. and **Zúniga, R.E.H.** (1998) *Evaluación crítica del impacto del Huracán Mitch en Nicaragua.* Coordinadora Civil para la Emergencia y la Reconstrucción (CCER), Managua.

Linz, J. and **Stepan, A.** (1996) *Problems of Democratic Transition and Consolidation.* The Johns Hopkins University Press, Baltimore.

Lipschutz, R. (1992) Restructuring world politics: the emergence of global civil society, *Millennium* **21**(3): 389–420.

MacDonald, L. (1994) Globalising civil society: interpreting international NGOs in Central America, *Millennium* **23**(2): 267–286.

—— (1997) *Supporting Civil Society: the Political Role of Non-Governmental Organisations in Central America.* Macmillan, Basingstoke.

McIlwaine, C. (1998a) Contesting civil society: reflections from El Salvador, *Third World Quarterly* **19**(4): 651–672.

—— (1998b) Civil society and development geography, *Progress in Human Geography* **22**(3): 415–424.

Olson, M. (1982) *The Rise and Decline of Nations*. MIT Press, Boston.

Puntos de Encuentro (1998) Los mecanismos para la movilización de recursos de las ONGs en Nicaragua, Fundación Puntos de Encuentro para Transformar la Vida Cotidiana, Managua.

Scott, C. and **Hopkins, R.** (1999) The economics of non-governmental organisations. The Development Economics Discussion Paper Series No. 15, STICERD, London School of Economics and Political Science, London.

Segundo Encuentro Regional (2000) *Memoria del segundo encuentro regional de la sociedad civil in Centroamérica 9,10,11 de Noviembre 1999*. Carqui Press, Managua.

Serbin A. (1998) Globalisation, democratic deficit and civil society in the Greater Caribbean integration process, in Wickham, P., Duncan, N., Wedderburn, J., Antrobus, P., Serbin, A., Jácome, F., Augustus, K. (eds) *Elements of Regional Integration: the Way Forward, Critical Issues in Caribbean Development*, No. 6. Ian Randle Publishers in association with the Caribbean Policy Development Centre, Kingston, Jamaica: 25–47.

Thomas, A. (1992) Non-governmental organisations and the limits to empowerment, in Wuyts, M., Mackintosh, M. and Hewitt, T. (eds) *Development Policy and Public Action*. Oxford University Press, Oxford: 117–146.

Toye, J. (1987) *Dilemmas of Development*. Blackwell, Oxford.

United Nations (1999) Partners in the reconstruction and transformation of Central America. Second Consultative Group Meeting for the Reconstruction and Transformation of Central America, Stockholm, Sweden, 25–28 May.

Vakil, A. (1997) Confronting the classification problem: toward a taxonomy of NGOs, *World Development* **25**(12): 2057–2070.

World Bank (1995) Co-operation between the World Bank and NGOs: FY 1994 Progress Report, Operations Policy Group and Operations Policy Department, February 1995.

CHAPTER 11

Conclusion

Cathy McIlwaine and Katie Willis

Just as Simon and Närman (1999: 273), in the first book in this series, shied away from 'the temptation to indulge in crystal gazing in the name of some profound fiftieth anniversary, end-of-century or millennial insight', we too are reluctant to prophesy the future of Middle America at the beginning of the twenty-first century. Instead, we would like to tie together some of the principal themes to emerge from this collection under the general rubric of 'challenges and change'. Important to highlight at the outset is that challenges and change are ambiguous notions that can be interpreted in a multitude of different ways. This makes them eminently suitable for thinking about the future of Middle America in that there are grounds for both pessimism and optimism.

Underlying this, and returning to the definitional issues with which we began the book, we would like to reiterate the utility of the term 'Middle America' to describe Mexico, Central America and the Caribbean. While the chapters have illustrated huge diversity within this broad geographical area, considerable economic, social, political and cultural commonalities have also emerged. For instance, while both Chant and Howard identify a host of differences among and within countries of Middle America in relation to gender and ethnic identities and household configurations, at the same time they outline some important axes of convergence and congruence. Thus, we feel that the term allows for greater specificity than the terms 'Latin America' and/or the 'Caribbean'. Furthermore, certain issues that are often made invisible when discussing the region as a whole, emerge forcefully when focusing on Middle America. An obvious example here is the issue of poverty; Middle America comprises some of the most impoverished nations of the world (such as Haiti, Nicaragua and Honduras), yet they are often overlooked when grouped as Latin America and the Caribbean, which is generally assumed to be a 'middle-income' region (see McIlwaine, this volume). Furthermore, Middle America as a whole tends to suffer from greater environmental vulnerabilities than its South American counterparts. This has been most obviously illustrated in the 1990s with reference to hurricanes 'Mitch' and 'George', and the Montserrat volcano among other disasters (see Bradshaw *et al.*, this volume on 'Mitch'). Indeed, as this conclusion is being written (in February 2001), El Salvador has just experienced its third earthquake within a month. Therefore, while careful to embrace the diversity of Middle America, as well as the inclusion of politically, and indeed developmentally 'unusual' countries such as Cuba, within its scope, we remain convinced that the notion of Middle America is extremely useful.

In the light of this, and returning to the issue of challenges and change, many Middle American societies have undergone significant transformations in the last few decades. While these may involve unanticipated events thrust upon nations, such as the natural disasters mentioned above, they also include the significant political violence and social unrest that have characterised Central America in particular. Both types of transformation have wrought undeniable hardship for the peoples of Middle America, with some sections of society, usually the poor, women and indigenous populations, experiencing the worst effects. However, changes can create both positive and negative ramifications and challenges for the future. Natural disasters such as earthquakes and hurricanes are obviously devastating in terms of loss of lives, severe disruptions to local communities, destruction of infrastructure and long-term consequences for economic growth (see Barton and Pelling, this volume). While these are undeniably deleterious, disasters can sometimes lead to an opening up of spaces of social and political transformation. Indeed, with reference to hurricane 'Mitch' in Nicaragua, Bradshaw *et al.* (this volume) highlight some significant transformations in state–civil society relations that were brought about by the disaster. Furthermore, the challenge for the future in Nicaragua and other countries in Middle America is to reduce the long-term vulnerabilities of the population in terms of poverty and infrastructure in particular; although the occurrence of natural disasters can rarely be accurately predicted, their effects are generally less severe in countries where people have access to safe and robust housing and infrastructure, and where governments can respond quickly, equitably and effectively with sufficient funds.

Different in nature and scope from natural disasters, the political conflicts that plagued Central America throughout the 1980s and 1990s also have ambiguous implications. While the loss of life and barbarity of these conflicts, especially in Guatemala and El Salvador, are difficult to justify, the peace accords signed in these two countries have attempted, to some extent, to address some of the deep-seated inequalities that led to the conflicts in the first instance. The extent to which this has been truly achieved is debateable. Nevertheless, a semblance of political stability has returned to these countries and some important achievements have been made. For example, in Guatemala, although discrimination against the indigenous population remains, incorporated within the Peace Accords of 1996 are the Indigenous Accords which, if lacking in concrete proposals, at least attempt to widen debates on how one of the largest indigenous populations in

Latin America can co-exist peacefully with their *ladino* counterparts (Howard, this volume). However, it is important to point out that the legacy of armed conflict in Guatemala and elsewhere in Middle America is profound. In particular, social networks and social capital have been severely undermined, communities have been ripped apart, and mistrust permeates many dimensions of social, political and cultural life (Moser and McIlwaine 2001 on Guatemala). Moreover, one of the most important challenges for the future in relation to conflict, is the ways in which political violence has given way to alarming levels of social and economic violence (ibid.). As pointed out in the introduction (Willis and McIlwaine), violence and crime are threatening to undermine political stability in the region in the twenty-first century; this obtains not only in Central America, but also in many Caribbean nations untouched by civil wars, especially in Jamaica, and even in Barbados (a country renowned for its high standards of living). Bound-up with this contemporary violence are poverty and social exclusion which are often cross-cut by ethnicity and gender (Howard and Pelling, this volume).

As well as these types of sudden and obvious change, Middle America has also experienced more gradual transformations in a number of areas, with similarly contradictory ramifications. Many of the latter are linked directly and indirectly with processes of globalisation. At the individual and family levels, Chant (this volume) highlights how household configurations have evolved and changed in Middle America. In particular, while emphasising household diversity, Chant identifies a broad trend towards the dissolution of the 'patriarchal family model' reflected in the increasing incidence of female household headship and, especially, lone motherhood, along with higher levels of divorce, declining rates of legal marriage and an increase in numbers of out-of-wedlock births. Partly linked with these patterns has been a gradual reorientation in the nature of gender identities and ideologies from those based on strict divisions between the male breadwinning role and female homemaking roles, towards a blurring of this divide. Enhanced access to paid employment among women has accounted for some of these changes. In turn, these are often linked with wider globalisation processes, especially related to employment opportunities that have opened up in export-oriented factories in Mexico, the Dominican Republic, Guatemala, Costa Rica and El Salvador among others (Willis, this volume). However, while these processes have led to some positive shifts towards more egalitarian gender divisions, they have also precipitated some negative implications in terms of the disintegration of household support networks, as well as a so-called 'crisis in masculinities' in which men's traditional gender roles have been questioned and threatened. In some places, the latter 'crisis' in particular has been linked with increased conjugal conflict and wider suggestions of societal breakdown. The challenges identified by Chant in these spheres revolve around the need to move away from idealising the two-parent patriarchal family, to accept household diversity and to embrace greater gender equality. It is also necessary to recognise ethnic diversity in family patterns in the region (Howard, this volume). Indeed, on an optimistic note, Chant outlines how seeds for such change have been sown in Costa Rica, a nation from which the rest of Middle America could possibly learn.

Also under the general rubric of gradual changes associated with globalisation has been the environmental sustainability of both rural and urban transformations. Thorpe and Bennett (this volume) have illustrated how, in rural areas, Middle America's heavy dependence on primary agricultural products has led to both economic growth and development for some (usually the small minority of wealthy elites), and economic vulnerability for others (usually the majority of peasantry and landless labourers). These contradictory effects of the evolution of agriculture in the region remain in contemporary Middle America in relation to the variegated nature of the environmental crisis or crises. Barton (this volume) points out that environmental degradation varies according to specific habitats, as well as particular socio-economic and political circumstances. None the less, he points out that many specific aspects of contemporary development linked with globalisation processes have had negative implications for the environment. In particular, he cites the shift towards non-traditional export products such as flowers, high-value fruit and vegetables, aquaculture and tourism as potentially extremely damaging for the environment, along with the effects of the expanding tourist sector in the region as a whole. Again, contradictions prevail. Agricultural transformations and tourism have saved some countries and areas from economic collapse (Willis, this volume), yet have also threatened fragile ecosystems. However, the promotion of ecotourism in many areas, while not always as environmentally friendly as purported, has managed to meet both economic growth and ecological requirements. The challenges highlighted by both Thorpe and Bennett and Barton thus primarily relate to the need for sustainable environmental management of traditional and non-traditional agricultural and tourist activities, as well as conservation initiatives for environmentally fragile ecosystems.

The need for sustainable management extends to cities in Middle America as urbanisation continues apace. As with the natural environment, urban growth has created problems of sustainability in relation to pressure on housing and urban services provision, as well as pollution, water supplies and sanitation facilities (Pelling, this volume). In some countries, urban sustainability has been further pressurised by concentrations of industrial development. Although industrial development is not always concentrated in urban areas, the vast majority of export-oriented manufacturing activities are located in towns and cities in Middle America. As central pillars of economic diversification intricately linked with globalisation processes, these industries not only have important environmental effects in terms of pollution, but also have ambiguous implications for the economy as a whole and the workforces employed there. As Willis (this volume) points out, these industries provide much-needed employment, yet do so on profoundly unequal bases, and often at the expense and exploitation of women who form the majority of labour forces. The challenge in this field, which is very difficult in the face of the movement of global capital, is to create sustainable employment opportunities that provide decent and non-exploitative wages in companies that are not entirely dependent on the pursuit of cheap labour and the provision of extensive tax, tariff and regulatory benefits on the part of Middle American governments.

These challenges are relevant for all forms of employment. Indeed, export-oriented factories provide only a minority of total employment in the region. Instead, informal sector employment is often the main form of livelihood, at least for the poor of the region (McIlwaine *et al.*, this volume). Although this type of work can provide significant opportunities for a minority, the bulk of people employed in these types of activity are barely making a living. Especially during the 1980s and 1990s, and related with neoliberal reforms and SAPs, informal employment expanded in most countries, both as a form of employment on its own account and as the informalisation of formal sector employment, especially export-manufacturing subcontracting.

Concomitant with neoliberal restructuring and underlying many other processes has been an increase in poverty, inequality and vulnerability in most countries of the region (McIlwaine, this volume). Although in many countries this relates to the adverse effects of globalisation processes and SAPs, the region has a long history of deep-seated inequality and social exclusion. Indeed, this has contributed in large part to the widespread exodus from the region in the form of international migration, especially to the US (Howard, this volume). However, within some Middle American countries, these inequalities are worsening rather than ameliorating. In Central America, for example, the poorest countries – Honduras and Nicaragua – have remained economically stagnant while the wealthier nations have witnessed significant economic growth (Esquivel *et al.* 2001). The challenges, recognised by international donor organisations and financial institutions, are to address these inequalities and reduce the debt burden of the poorest countries. Some efforts have been made towards reaching this goal through the HIPC initiative that has included both Nicaragua and potentially Honduras – the only two Latin American or Caribbean countries considered (McIlwaine, this volume). Nevertheless, it is also important to emphasise that economic growth and recovery on their own are not sufficient for sustainable development to be achieved in the region.

As we have tried to show throughout this collection, socio-economic, political, cultural and environmentally aware development are also necessary components of sustainable change in Middle America. Bearing in mind that favourable transformations will be gradual, the challenges for the future are to ensure that the positive effects of globalisation processes are enhanced and the negative ramifications minimised. Indeed, this echoes the key message of the UK government's recent White Paper on International Development that calls for 'globalisation in the interests of poor people, creating faster progress towards International Development Targets' (DFID 2000: 4) that include poverty reduction, gender equality and improved access to health and education services.[1] While these are global

---

[1.] These targets include reducing by one half the proportion of people living in extreme poverty by 2015, universal primary education by 2015, progress towards gender equality and empowerment, a reduction by two-thirds in infant mortality rates, access to reproductive health care systems, and finally, the implementation of national strategies for sustainable development by 2015 to ensure that current patterns of the loss of environmental resources are reversed both globally and nationally (DFID 2000: 3).

targets, there are certain challenges specific to Middle America. These include maintaining peace and fragile democratisation efforts, keeping crime and violence under control, reducing social exclusion and inequality (especially on gender and ethnic grounds), and making the region less vulnerable to natural hazards. Some Middle American countries have already been successful in achieving some of these aims. To conclude on a positive note, countries such as Costa Rica and Barbados that have already made significant strides towards sustainable development, might act as beacons of hope for those less fortunate in the region. Indeed, they also illustrate that solutions to the challenges outlined above are achievable in practice as well as in theory.

References

Department for International Development (DFID) (2000) *Making Globalisation Work for the World's Poor: an Introduction to the UK Government's White Paper on International Development.* DFID, London.

Esquivel, G., Larraín, F. and **Sachs, J.D.** (2001) Central America's foreign debt burden and the HIPC initiative, *Bulletin of Latin American Research* **20**(1): 1–28.

Moser, C. and **McIlwaine, C.** (2001) *Violence in a Post-Conflict Context: Urban Poor Perceptions from Guatemala.* World Bank, Washington, D.C.

Simon, D. and **Närman, A.** (1999) Conclusions and prospects, in Simon, D. and Närman, A. (eds) *Development as Theory and Practice: Current Perspectives on Development and Development Co-operation.* Addison Wesley Longman, Harlow: 269–275.

Index